Changing Precipitation Regimes and Terrestrial Ecosystems

Changing Precipitation Regimes and Terrestrial Ecosystems

A North American Perspective

Jake F. Weltzin & Guy R. McPherson,

EDITORS

The University of Arizona Press   Tucson

The University of Arizona Press
© 2003 Arizona Board of Regents
First Printing
All rights reserved
⊗ This book is printed on acid-free, archival-quality paper.
Manufactured in the United States of America

08   07   06   05   04   03   6   5   4   3   2   1

Library of Congress Cataloging-in-Publication Data
Changing precipitation regimes and terrestrial ecosystems : a North American perspective /
Jake F. Weltzin and Guy R. McPherson, editors.
p. cm.
Includes bibliographical references (p.   ).
ISBN 0-8165-2247-2 (cloth : alk. paper)
1. Plant-water relationships. 2. Plants, Effect of soil moisture on. 3. Vegetation and climate.
4. Precipitation (Meteorology). 5. Ecosystem management. I. Weltzin, Jake F.
II. McPherson, Guy R. (Guy Randall), 1960–
QK870 .C484 2003
581.7—dc21
2002008738

British Library Cataloguing-in-Publication Data
A catalogue record for this book is available from the British Library.

To the memory of Joy Belsky
—pioneering scientist, courageous
conservation biologist, and inspiring scholar

Contents

Foreword

JAMES H. BROWN

After growing up in Illinois, author Mary Austin spent most of her life in the intermountain basins and deserts of California, Arizona, and New Mexico. In *The Land of Little Rain* (1903) she paints a picture of rugged, eroded landscapes, punctuated with sparse shrubby and spiny plants. Although she was not a scientist, Austin's graphic descriptions capture the nature of the vegetation and the central role of water in a region where it is scarce.

> [T]he desert begins with the creosote. This immortal shrub spreads down into Death Valley and up to the lower timber-line, odorous and medicinal as you might guess from the name, wandlike, with shining fretted foliage. . . . There is neither poverty of soil nor species to account for the sparseness of desert growth, but simply that each plant requires more room. So much earth must be pre-ëmpted to extract so much moisture. The real struggle for existence, the real brain of the plant, is underground; above there is room for a rounded perfect growth. (Pp. 10–12)

The authors of the chapters in the present volume are scientists, but they, too, paint a picture of the dependence of plants on water, of vegetation on precipitation. Mary Austin refers to the "real brain" of plants as being underground—i.e., in the roots. Her metaphor is appropriate. The scientists in this book write about "rooting strategies," the elaborate evolutionary adaptations and facultative changes that enable plants to gain access to belowground resources.

One might think that the relationship between precipitation and vegetation would be so well worked that it does not warrant a book-length review. Nothing could be further from the truth. There has indeed been a great deal of research on plant-water relations at all scales, from the water economy of individual plants to the water fluxes through terrestrial ecosystems. At all scales, however, there is much still to be learned. More importantly, there is a need to integrate information across scales: to understand how ecosystem responses depend on plant performance, and how water relations vary across space and time among plants of the same and different species.

In this book the experts address these questions. The emphasis is on the insights that come from comparisons within and among ecosystems in temperate North America. The authors of various chapters report results of field studies in Sonoran and Chihuahuan desert, sagebrush steppe, oak savanna, tall- and mixed-grass prairie, and deciduous forest. Equally important, the authors report a variety of insights and results that come from different approaches and methodologies. These include spatial comparisons of vegetation structure and function across ecosystems that differ in soils and climate; intensive analyses of changes in plant architecture and physiology in response to temporal variation in precipitation within ecosystems; experimental studies that use rainout shelters and other techniques to manipulate water availability; and modeling approaches that characterize the relationships between climatic variables and vegetation types. A common goal of these studies is to assess vegetation response to major shifts in climate that appear to be occurring at present and may be the norm in the future. Much of the recent emphasis has been on changing temperature patterns due to global warming. The emphasis of this volume is on changes in precipitation regimes. These may be due in part to global warming but also to other factors, such as changes in atmospheric circulation and land surfaces.

Of course this book cannot answer all of the questions. Its strength and its limitations are due to its restricted subject matter. The book focuses on the influence of precipitation on only one trophic level, the primary producers. Effects of precipitation on consumers and decomposers, and potentially important indirect effects of climate on vegetation through these other trophic levels, are not considered. The book also focuses exclusively on temperate North American ecosystems. In doing so, it complements other recent studies of ecosystems with similar and contrasting climates outside the United States.

On the other hand, the restricted subject matter and careful editing make for a compact, focused, and timely volume. It is an authoritative work— a summary of the current state of a rapidly changing science by many of the leading scientists in the field. But its value lies not so much in its factual materials and literature reviews as in its point of view, not so much in answering the questions as in asking how they should be asked—and answered. The book asks how terrestrial vegetation responds to varying precipitation regimes—both the "natural" range of variation that plants have experienced in the past and are adapted to, and the unusual new variation that is due to activities of modern humans. It makes clear that the methods for answering this question have changed dramatically in just the last few years.

The message that I take away is that both the temporal and spatial patterns of environmental variation and the nature of the vegetation responses are complex. Students and professionals, managers and policymakers will need to account for this complexity. Anthropogenic climatic changes and other human impacts are altering species composition, vegetation structure, and ecosystem processes in terrestrial communities. These changes have important implications for the survival of endemic species, the management of grasslands and forests, and the productivity of agroecosystems. The topics addressed in this book are central to all of these issues.

Changing Precipitation Regimes and Terrestrial Ecosystems

I

Predicting the Response of Terrestrial Ecosystems to Potential Changes in Precipitation Regimes

JAKE F. WELTZIN & GUY R. MCPHERSON

Human-induced change in global environments is one of the most important and timely topics facing society. As the effects of human activities on Earth's climate, sea levels, and ecosystems become more apparent in the coming decades, global change issues likely will become even more important to global citizens and their governmental representatives.

One important aspect of global change is the potential response of terrestrial ecosystems to changing environmental conditions. Anthropogenic increases in atmospheric carbon dioxide concentration have both direct and indirect ramifications for natural ecosystems: global increases in carbon dioxide may stimulate plant growth, but they will also increase surface temperatures and change precipitation regimes (Houghton et al. 2001). Considerable research has described the effects of increasing atmospheric carbon dioxide concentration and expected increases in temperature on ecosystems (e.g., Koch and Mooney 1996; Harte and Shaw 1995), but little research has focused on changes in the amount or seasonality of precipitation anticipated in the next few decades.

Increases in concentration of atmospheric carbon dioxide are expected to increase global temperatures and thereby alter the amount, seasonality, and intensity of precipitation on global and regional scales (Houghton et al. 2001). Predictions of future precipitation regimes in 2030 and 2095 from two general circulation models (GCMs)—the Canadian Climate Center (CGCM1) and the Hadley Centre in the United Kingdom (HADCM2)—were used for the United

States national assessment (National Assessment Synthesis Team 2000). The CGCM1 model predicts significant reductions in summer and winter precipitation in the Southeast and Great Plains regions by 2095. In contrast, HADCM2 simulations for 2095 show increased precipitation throughout most of the United States, except for summer reductions in the Southwest. More recent model runs using two Vegetation/Ecosystem Modeling and Analysis Project (VEMAP) models—the Mapped Atmosphere-Plant-Soil System (MAPSS) and MC1 (Daly et al. 2000)—suggest that increased warming alone could increase evaporative demand and increase droughts in the Southeast, southern Rockies, parts of the Northwest, and the Gulf Coast even though annual precipitation might increase. Moreover, the frequency of extreme precipitation events may increase as a result of the intensification of the hydrologic cycle: in a recent review, Easterling et al. (2000) noted that single- and multi-day heavy precipitation events around the globe became more frequent during the twentieth century, and that areas of the world affected either by drought or excessive wetness have also increased.

There is growing scientific awareness of how changes in precipitation regimes within the context of global change may affect terrestrial ecological systems. Changes in global and regional precipitation regimes are expected to have important ramifications for the distribution, structure, composition, and diversity of plant communities (e.g., VEMAP Members 1995; Neilson and Drapek 1998; Bachelet and Neilson 2000). For example, a recent comparison of two models designed to simulate time-dependent and equilibrium changes in the distribution of potential vegetation indicated spatial shifts in major vegetation types on regional scales, with attendant ramifications for carbon sequestration (Bachelet et al. 2001). Similarly, changes in the intensity, frequency, and seasonality of precipitation may affect rates of production and decomposition, biogeochemical cycling, frequency of wildfire, availability of water resources in terrestrial ecosystems, and feedback between the biosphere and the atmosphere (Melillo et al. 1996; Grime et al. 2000; Joyce and Birdsey 2000; Easterling et al. 2000; Bachelet et al. 2001). Changes in precipitation also may increase the susceptibility of ecosystems to invasion by nonnative plant species (Dukes and Mooney 1999; Smith et al. 2000), and may affect spatial and temporal dynamics of consumers dependent upon resources at lower trophic levels (Ernest et al. 2000). Further, changes in precipitation that affect the distribution and abundance of plants may have ramifications for the conservation of species of biological or social concern.

In short, scientific understanding and effective management of plant species and communities in the face of climate change will depend on our ability to accurately predict their response to different biotic and abiotic driving variables. This in turn will depend on a mechanistic understanding of individual and combined species response to resource limitations under changing environments. To this end, several large-scale field experiments have been designed to assess the physical and biological mechanisms that may control the effects of changes in precipitation regimes on individual plants, plant populations, and plant communities and their ecosystems. However, in contrast with carbon dioxide and surface temperature research—the sole focus of many books, journals, and scientific meetings—there has been no central forum for discussion of information about this newly breaking arena of global change research.

Therefore, we organized a symposium at the 1998 Annual Meeting of the Ecological Society of America designed to increase awareness that (1) anthropogenic climate change will include shifts in the amount, seasonality, and intensity of precipitation, with attendant ramifications for ecosystem structure and function, and that (2) field experiments represent an efficient method to determine effects of potential precipitation regimes on ecosystems at landscape to regional scales. To our knowledge, this symposium was the first such gathering of ecologists concerned with this important, yet little-studied, component of global and regional climate change. At that time, we began to develop these ideas more fully in the form of an edited monograph encompassing the most important aspects of interactions between global change, precipitation, and terrestrial ecosystems. This volume is one result, and it has the following purposes: (1) to explore various approaches that can be used to predict responses of ecosystems to changes in precipitation regimes, (2) to bring together disparate components of this particular aspect of global change research, and (3) to highlight the emerging research interest in the importance of precipitation regimes in structuring natural ecosystems.

There is no dearth of research on the role of precipitation in structuring terrestrial plant communities and ecosystems, especially within the last half of the twentieth century. Much of this research has been based on observational or correlational approaches, which are limited in their ability to separate the relative importance of the many confounding factors that may change in association with precipitation. Analyses of long-term records of precipitation vis-à-vis various attributes of plant populations or communities have provided a strong basis for understanding system response to spatial and temporal variability in soil moisture. However, it is sometimes difficult to distinguish the role of other factors that may vary independent of precipitation events or regimes (e.g., changes in temperature, insect outbreaks), especially in the face of stochastic precipitation events that may structure communities and ecosystems.

Experimental simulation of soil moisture is another approach to determining how precipitation may affect communities and ecosystems. Although several techniques can be used to manipulate soil moisture (see Owens, chapter 5, this volume), they face a series of conceptual and logistical challenges. Foremost, in contrast to other biogeochemical aspects of global change (e.g., concentrations of radiatively active gases in the atmosphere and surface temperatures), there is little agreement regarding predictions about the magnitude and direction of future changes in precipitation regimes (VEMAP Members 1995; Mahlman 1997; Schimel et al. 2000). Discrepancies between model predictions depend in part on the spatial scale (i.e., cell size) of the model coupled with the effects of regional topography. Most GCMs do not have sufficient resolution to make concrete predictions as to how precipitation will change, particularly in topographically complex regions such as the southwestern United States. For example, the HADCM2 GCM developed by the Hadley Centre at the U.K. Meteorological Office predicts that the Southwest will experience drier summers and wetter winters by the year 2030 (VEMAP Members 1995). In contrast, a nested regional climate model (RegCM) developed by the National Center for Atmospheric Research (NCAR) predicted that a doubling of current carbon dioxide (CO_2) concentration will decrease the amount of winter precipitation in the southwestern United States (Giorgi et al. 1998).

Further, logistical constraints to the experimental application of precipitation are many, and include issues related to method and timing of application, management of water on the surface and in the soil of field plots, access to field plots during rainfall events or rainy seasons, and undesired experimental artifacts (e.g., herbivory, shading of plots by infrastructure, alteration of microclimate), not to mention the sheer mass of large volumes of water. For example, the experimental addition of small amounts of water (in many cases corresponding to small but significant precipitation events) may produce no response because of hot, dry conditions when the water is applied. In contrast, relatively small plots—surrounded by a matrix of natural vegetation—that receive large experimental additions of water may attract herbivores that negate the effects of water augmentation. Although such constraints may be overcome by the careful design of experiments, monetary limitations often remain a daunting obstacle.

Despite these challenges, we believe the future is bright for experimental research on altered precipitation regimes. Likely future scenarios generated by regional climate models or GCMs provide reasonable bounds on experimental manipulations, and ecologists have always relished challenges associated with field research. Recent articles on precipitation-manipulation experiments attest to the growing interest in the topic and the ability of researchers to overcome conceptual and logistical difficulties (e.g., Gunderson et al. 1998; Reynolds et al. 1999; Grime et al. 2000; Hanson 2000; Weltzin and McPherson 2000).

The chapters in this volume are arranged first to elucidate the importance of precipitation regimes, soil characteristics, and soil moisture to the distribution and abundance of plants and vegetation, and then to provide specific information about recent and ongoing investigations of the response of plant populations, communities, and ecosystems to manipulation of soil moisture in a variety of field settings. After this introductory chapter, chapters 2 and 3 outline the role of geological substrate and the functional architecture of plant roots in dictating the response of plants to current and potential future precipitation regimes. These chapters provide the context for interpretation of plant response to changes in available soil moisture caused by natural variation or experimental manipulation. Chapter 4 describes the use of a well-known and broadly tested biogeographic model (MAPSS) to (1) explain apparent anomalies in the distribution of vegetation types on regional scales vis-à-vis broad-scale patterns of atmospheric circulation and precipitation amount and seasonality, and (2) predict the response of plant populations and communities to potential changes in precipitation at landscape to regional scales. Chapter 5 provides an overview of field methods used to manipulate soil moisture and simulate changes in precipitation regimes. This chapter is particularly useful from an applied perspective, because experimental techniques of precipitation manipulation, though diverse, are still being developed, and actual implementation can be a daunting task.

Chapters 6–10 contain a series of five case studies that investigate the role of precipitation amount, seasonality, and frequency in shaping terrestrial ecosystems of the western and central United States, where precipitation already constrains community and ecosystem structure and function, and where the majority of the world's experimental research on this topic is being conducted. The case studies range from sagebrush steppe (with total annual precipitation of 300 mm) to eastern deciduous forest (which receives 1350 mm each year). As chapters, the case studies are arranged along a gradient of increasing total annual precipitation, although as highlighted in several chapters the intraseasonal patterns of precipitation are often more important than annual amounts of precipitation. Nonetheless, this particular organization serves to highlight the differential response of ecological systems to changes in seasonal *and* annual precipitation totals.

The case studies in chapters 6–10 highlight the important role of experiments in elucidating relationships between precipitation, soils, and plants. Results of the case studies indicate surprising responses not likely predicted from the historically common descriptive approach to assessing vegetation response to changes in environmental driving variables. They also illustrate the importance of the many interacting factors that constrain the response of plants—both individuals and populations—to changes in precipitation regimes. In the final chapter, we briefly synthesize the text and describe future research needs.

This book is targeted at ecologists, global change scientists, climatologists, natural resource managers, social scientists, and students who are interested in the effects of precipitation on ecosystems. It is not specifically intended to be used as a course textbook, but it may prove useful as a supplementary text for courses in global change, hydroclimatology, meteorology, natural resources, or ecology. In addition, it may serve as a primary text for a seminar or colloquium dedicated to the study of this aspect of global change. Upper-division undergraduate students and graduate students with rudimentary backgrounds in ecology should have no difficulty reading and understanding the book.

Acknowledgments
We would like to thank Linda McMillan, Melissa Moriarty, and Charissa Nadeau for their help in assembling the final text. We also extend thanks to each of the contributors for their patience and willingness to share their insights and research findings. Comments from two anonymous reviewers and each of the chapter contributors greatly improved the structure, content, and synthesis of the text.

2

The Interface between Precipitation and Vegetation

The Importance of Soils in Arid and Semiarid Environments

JOSEPH R. MCAULIFFE

As described in chapter 1, changes in global climate over the next century are likely to include altered precipitation regimes. Shifts in the amount, frequency, or seasonality of precipitation that accompany global climate change are likely to differ among various regions of the Earth's surface. In addition, their impact on the structure and function of ecological systems will be system-dependent. For example, in arid and semiarid environments, even slight changes in precipitation regimes may produce large ecological effects. Whether climate changes of the future are "natural" or anthropogenic, they have the potential to greatly impact human well-being, because arid and semiarid lands cover about a third of the Earth's land surface and support a fifth of the world's human population.

The scarcity of water limits most ecosystem processes in arid and semiarid regions (Noy-Meir 1973; Sala et al. 1997; Reynolds et al. 1999). Before water from precipitation becomes available to plants, it must first enter and be stored in the soil. Different soil conditions can create diverse soil moisture environments in an area that receives a given amount of precipitation. This soil hydrologic diversity in turn generates considerable diversity and complexity in ecological responses (McAuliffe 1999b). The purpose of this chapter is to outline how soil conditions mediate vegetation responses in water-limited regions. Understanding relationships between soil, soil hydrologic behavior, and plant response is a necessary step to predicting the kinds and magnitudes of vegetation change that may occur in response to any kind of environmental perturbation, including climate change. The chapter concludes with a discussion of the types of information, investigations, and collaborations that can contribute significantly to the prediction of ecological responses to future climates in arid and semiarid regions.

Spatial and Temporal Distributions of Soil Moisture

Three characteristics of the spatial and temporal distributions of soil moisture strongly influence vegetation composition, structure, and productivity in arid and semiarid regions: vertical (i.e., depth) distribution of soil moisture, localized evenness versus patchiness in the horizontal distribution of soil moisture, and temporal persistence of soil moisture.

Depth Distribution

Attributes of precipitation events (amount, intensity, frequency, and seasonality) together with soil characteristics influence the amount of precipitation that infiltrates the soil surface and the depth to which it percolates. In arid and semiarid regions, most precipitation events are small and only wet the soil shallowly (Sala and Lauenroth 1982); less frequent, larger precipitation events may penetrate to deeper layers of the soil.

Rainfall intensity (often related to season) is a strong determinant of the amount of moisture that is absorbed by the soil versus lost as runoff. In the American Southwest, a much greater proportion of the precipitation delivered by intense summer convective storms is lost as runoff in comparison with the gentler precipitation of winter frontal storms. For example, in the Walnut Gulch Experimental Watershed of southeastern Arizona, approximately 70 percent of the year's precipitation is delivered in summer convective storms, but 95 percent of all runoff for the year is derived from these storms (Osborn 1983).

Soil characteristics, especially soil texture, further control surface infiltration and the depth to which water percolates. Coarse-textured (i.e., sandy) soils typically foster more rapid and deeper movement of water than do fine-textured soils (Walter 1979; Noy-Meir 1973; Sala et al. 1997). For example, at the Jornada Experimental Range in southern New Mexico, Reynolds et al. (1999) showed that even above-average amounts of summer precipitation did not recharge soil water at depths of 30 cm or more in sandy, loam-textured soils ("lower bajada" site). However, summer precipitation produced slight recharges at a 30 cm depth in slightly coarser-textured loamy sands ("upper bajada" site) and considerable recharge at depths of 30 cm and more in even coarser sands ("dune" site).

Horizontal Evenness

Soil characteristics further influence the spatial heterogeneity of soil moisture. Depending on certain soil features and/or the nature of the underlying parent materials, soil moisture at any particular depth may range from relatively even to extremely patchy. Deep soils formed in relatively fine, granular parent materials, such as gravelly alluvium, foster a relatively even horizontal distribution of soil moisture at any particular depth, although the vertical (depth) distribution of moisture may vary according to texture and horizon development (fig. 2.1A,B). In contrast, extremely rocky or similarly impenetrable substrates foster heterogeneous horizontal distributions of soil moisture. For example, where weathered and fractured bedrock or other similar substrates are located beneath relatively thin, weakly developed soils, narrow spaces of fractures and joints in the underlying bedrock are conduits for relatively deep but spatially heterogeneous recharge of moisture (Sternberg et al. 1996; fig. 2.1C). Similar patchiness may occur in soils with shallow, strongly cemented calcic horizons ("caliche," or "calcrete"), where plant roots penetrate cracks within these otherwise hard and impenetrable layers. Patchy vegetation cover and associated activities of burrowing animals may also contribute to horizontal unevenness in the distribution of moisture and other resources (Schlesinger et al. 1990; Whitford and Fenton 1999).

Temporal Persistence

The duration of plant-available soil moisture is in large part a function of the depth of moisture storage (see "Depth Distribution" above). Although shallow parts of the soil are wetted more often than are deeper layers, shallow soil moisture is more rapidly lost to evaporation. Consequently, the amounts of moisture stored deeper in the soil are more constant over time than the amounts stored nearer the surface (Noy-Meir 1973; Monson and Smith 1982; Schlesinger et al. 1987; fig. 2.2).

Deeper percolation and storage of moisture in a coarse-textured soil provide a more temporally persistent source of soil moisture for plants than the same amount of precipitation delivered to a finer-textured soil where the depth of percolation is considerably less. The temporal persistence of

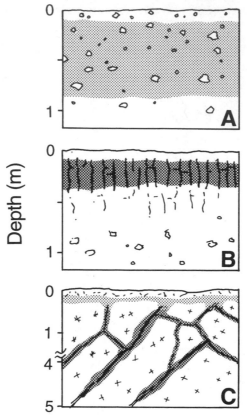

Figure 2.1. Vertical (depth) and horizontal distributions of moisture in contrasting soils and underlying parent materials following a moderate rain after the onset of a drying cycle. Shaded areas indicate distribution of plant-available water; darkness of shading indicates amount of water (light = low, dark = high). (A) Deep, coarse-textured soil on relatively fine, granular parent materials fosters rapid surface infiltration and deep percolation of moisture. (B) High moisture-holding capacity of a clay-enriched argillic horizon inhibits deep percolation. Shallow storage of water leads to extreme interseasonal variation in plant-available moisture. (C) Highly weathered bedrock below thin, weakly developed soils often promotes deep, horizontally uneven percolation along fractures and joint faces; note difference in vertical scale.

plant-available soil moisture may also be strongly influenced by subsoil features that concentrate and channel the downward movement of moisture, including fractures in weathered bedrock underlying thin soils (Sternberg et al. 1996) and changes in bulk density and creation of soil macropores due to burrowing activities of animals (Whitford and Fenton 1999). Although the horizontal distribution of moisture at any depth in such substrates may be highly variable, the concentration of moisture in limited zones (e.g., in deep

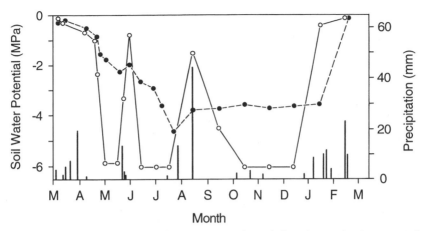

Figure 2.2. Seasonal variation in soil water potentials in shallow (10 cm depth; open circles, solid line) versus deeper (40 cm depth; closed circles, broken line) soil levels at a site in the Sonoran Desert near Phoenix, Arizona. Amounts and timing of individual precipitation events are shown by the vertical lines. Adapted from Monson and Smith (1982).

vertical fractures) may foster the movement of water to considerable depth, where it is largely insulated from evaporative losses to the atmosphere (fig. 2.1C).

In arid and semiarid environments, soil conditions, such as clay-rich argillic horizons, that inhibit deep percolation of water and limit moisture storage to shallow depths *amplify* the seasonal variability in plant-available moisture. In contrast, soil conditions that facilitate deeper movement and storage of moisture *dampen* seasonal variation in soil moisture supply (Noy-Meir 1973; Burgess 1995; McAuliffe 1999b).

Water Use by Plants

Plants of arid and semiarid environments have diverse morphologies, phenologies, and physiologies. Stem and leaf succulents, drought-deciduous shrubs, drought-enduring evergreen shrubs, small trees (both evergreen and winter deciduous), perennial grasses, and ephemerals are some of the principal plant life forms that occupy water-limited environments. Despite this great diversity of life forms, the manner in which water is extracted from the soil by different kinds of plants can be arranged along a continuum ranging from *intensive* to *extensive* exploitation (fig. 2.3; Burgess 1995).

Intensive exploiters have dense, shallow root systems. They grow rapidly during the brief intervals when shallow soil moisture is available and

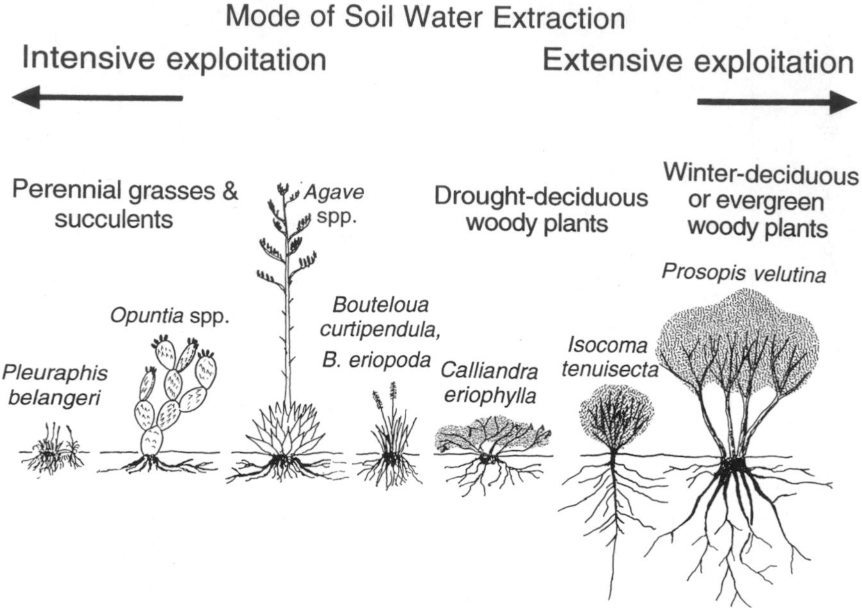

Figure 2.3. Modes of water extraction of plants common to warm, semiarid regions, illustrated by common plants found in semidesert grasslands in the American Southwest.

typically become dormant when these shallow stores of moisture become exhausted. Perennial warm season C_4 grasses such as curly mesquite *(Pleuraphis [= Hilaria] belangeri)* and tobosa *(P. mutica)* of warm semidesert grasslands and blue grama *(Bouteloua gracilis)* of shortgrass prairie are intensive exploiters of sporadic summer precipitation that only wets shallow layers of the soil (Sala and Lauenroth 1982, 1985; Reynolds et al. 1999). Many designations of plant life forms and plant functional types of arid regions recognize a dichotomy between the rooting depths of shallow rooted grasses and deeply rooted woody plants (Sala et al. 1997). However, many subshrubs such as snakeweed *(Gutierrezia* spp.) and some small drought-deciduous shrubs like false mesquite *(Calliandra eriophylla)* are positioned toward the intensive exploitation end of the water-acquisition continuum. The shallow root systems of these small woody plants often occupy the same, upper part of the soil as do those of grasses and similarly rely on a temporally variable supply of water. Drought-deciduousness allows these plants to persist in a dormant state during periods when shallow soil moisture is no longer available. Ephemeral plants also are intensive exploiters of shallow soil moisture; they survive lengthy dry periods as dormant seeds.

Water-storing plants may be thought of as a special type of intensive exploiter. The root systems of many succulents, including cacti, agaves, and yuccas, are shallow and limit these plants to brief bouts of water uptake before shallow soil moisture is exhausted. However, instead of becoming dormant during droughts, these plants rely on water stored within stems, leaves, or roots to carry on photosynthetic activity past the time when their roots can no longer extract moisture from the soil (Gibson and Nobel 1986).

Toward the other end of the water-use continuum, extensive exploiters are typified by larger woody plants that have both wide-spreading and deep roots. Access to more constant, deep moisture supplies allows extensive exploiters to maintain transpiration-demanding leaf function and even flowering during extended rainless periods. For example, in the Sonoran and Mohave Deserts and semidesert grasslands, cat-claw acacia (*Acacia greggii*) comes into leaf and flower during the driest part of the year in late spring, relying on soil moisture stored deep in the soil during the previous winter season. Other large woody plant species of the region, including velvet mesquite (*Prosopis velutina*) exhibit similar phenological patterns that indicate a reliance on the more temporally constant supplies of deep soil moisture (Burgess 1995; McAuliffe 1995; Reynolds et al. 1999).

For extensive exploiters, the soil depth from which water is extracted shifts considerably during plant growth and development. For example, the seedlings and juvenile stages of many large woody plants have only shallow roots, but these eventually develop into deeper and wider-spreading, extensive root systems (Harrington 1991; Donovan and Ehleringer 1992; Weltzin and McPherson 1997). Consequently, seedlings and very young woody plants require a dependable, uninterrupted supply of soil moisture obtained from the same shallow rooting zone occupied by intensive exploiters such as perennial grasses. This dependence on shallow soil moisture is an extremely precarious life history stage for young woody plants of arid and semiarid regions. These plants typically do not possess a capacity for drought dormancy as exhibited by true intensive exploiters. Woody plant survival during this critical developmental stage (often the first summer season of growth) depends on climatic, soil, and other ecological conditions that foster a relatively steady supply of moisture.

Vegetation Responses

Various spatial and temporal distributions of soil moisture availability (fig. 2.1) foster the predominance of plants that differ in their modes of soil moisture acquisition. The characteristic hydrologic behaviors as determined by soil properties may in some cases almost entirely preclude occupation by plants representing one end or another of the intensive-extensive exploitation continuum (fig. 2.3).

Plants capable of intensively exploiting the temporally inconstant supplies of shallow soil moisture predominate on soils containing well-developed argillic horizons that limit deep soil recharge (fig. 2.1B). Although downward penetration of moisture may be restricted, horizontal distribution of soil moisture may be relatively even, allowing uptake by relatively shallow, dense, fibrous root systems of perennial, warm-season grasses and other intensive exploiters. On the other end of the spectrum, thin, weakly developed soils on highly fractured bedrock substrates may contain deep moisture supplies that are extremely uneven in horizontal distribution, but more constant over time, fostering predominance of woody plants with extensive, deep root systems that can access these widely dispersed moisture sources (McAuliffe 1994, 1999b).

Some edaphic settings provide hydrologic environments that are capable of supporting both intensive and extensive exploiters. Deep, coarse-textured soils that readily transmit the precipitation signal and recharge both surface and deeper layers are capable of supporting both shallow-rooted intensive exploiters and more deeply rooted extensive exploiters (Sala et al. 1997). However, the predominance of one kind of plant or the other (e.g., perennial grasses versus woody plants) often depends on a variety of factors, including the reduction of the competitive ability of grasses by livestock grazing and changes in fire frequency (Glendening and Paulsen 1955; McPherson 1995; Scholes and Archer 1997).

The effects of contrasting soil hydrologies on the composition of vegetation are evident at spatial scales ranging from regional to local. For example, on a regional scale, two contrasting vegetation types, warm semidesert grassland and interior chaparral, broadly overlap in southern Arizona with respect to elevational distribution and average annual precipitation (semidesert grassland: 1100–1400 m elevation, 250–450 mm precip.; interior chaparral: 1050–1850 m elevation, 350–635 mm precip.; Brown 1982; Pase and Brown 1982). In Arizona, these two vegetation types flank the northern and eastern boundaries of more arid Sonoran Desertscrub located at lower elevations to the southwest (fig. 2.4). Despite the overlap in elevational range, chaparral and grassland are not evenly distributed across this broad zone surrounding the

Sonoran Desert. Semidesert grasslands predominate to the east whereas interior chaparral occurs to the north and west. Although there is a gradient of reduced summer precipitation in Arizona from east to west which might be invoked as the cause for the westward-increasing predominance of chaparral, the region in which chaparral predominates nevertheless contains large inclusions of semidesert grassland. Conversely, extensive regions of chaparral are found much further to the east in the Chihuahuan Desert region of southern New Mexico, Texas, and north-central Mexico (Pase and Brown 1982). This more easterly region typically receives larger and more predictable amounts of summer precipitation than the grasslands of southeastern Arizona, yet it is dominated by C_3 woody plants rather than C_4 perennial grasses. Clearly, regional distribution of chaparral versus grassland is not regulated by climate alone.

This biogeographic pattern is largely due to differences in lithology and topography, which yield highly contrasting kinds of soils and associated soil hydrologies. In Arizona, most chaparral occurs in the geological transition zone that separates the Basin and Range physiographic province in the south and southwest from the Colorado Plateau to the northeast. This transition zone is characterized by rugged, low mountains composed mostly of coarse-grained Precambrian plutonic and metamorphic rocks (fig. 2.4). Throughout much of the transition zone, these lithologies often form relatively unstable slopes that do not foster the development of well-formed soils. Coarse-grained plutonic (granitic) lithologies underlie more than half of the area covered by interior chaparral in Arizona; schist is the next most common lithology (Carmichael et al. 1978). Although soil horizonation is usually weak in these parent materials and topographic settings, these types of rocks are usually deeply weathered (Bull 1991) and permit extensive penetration by water along fractures and joints (fig. 2.1C). Chaparral is also found on other lithologies where soil formation is weak, but where water may penetrate the weathered, rocky substrate to great depths.

Extensive taproots of evergreen, sclerophyllous chaparral shrubs penetrate and extract water from these spatially heterogeneous, deep moisture supplies. Pase and Johnson (1968) reported 13 g of chaparral shrub roots per cubic foot (= 459 g/m^3) at a depth of 3.65 m at a site in central Arizona. In a study of chaparral in southern California, roots of shrubs also penetrated deeply ($\geq$ 4 m) along joint faces in highly weathered granitic bedrock (Sternberg et al. 1996). That study showed that during the summer dry

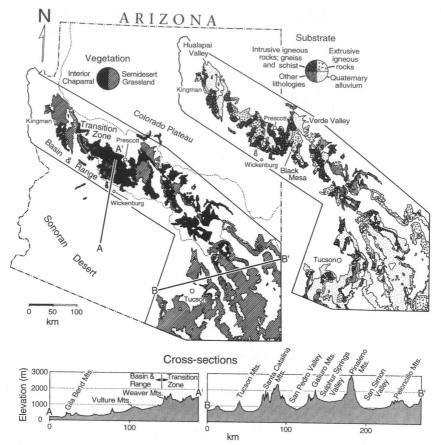

Figure 2.4. Geographic distribution of interior chaparral and semidesert grassland in Arizona (upper left) and the geological substrates underlying these two vegetation types (upper right). Locations of the cross-sections (A-A′, B-B′) are shown by the labeled lines on the map of Arizona.

season, shrub roots that occupied a 290 cm thick zone of weathered, fractured bedrock extracted nearly ten times the moisture than did roots occupying the overlying, 35 cm thick soil.

The inclusions of grasslands within the broad region occupied by interior chaparral in Arizona occur on different geological substrates and associated soils than those that support chaparral (fig. 2.4). Grasslands are principally found on either large, flat, mesa-like remnants of Tertiary basalt flows or Quaternary alluvium of basins (e.g., Black Mesa and the Verde and Hualapai Valleys; fig. 2.4). The massive, erosion-resistant basalts form geomorphically stable terrains that have fostered the formation of very old, well-developed soils with thick, clay-enriched argillic horizons. Soils formed in Quaternary alluvium vary considerably, depending in part on the age of the deposit (discussed more fully below).

Quaternary alluvial deposits also form the principal substrate that supports the most extensive areas of semidesert grasslands in the southeastern corner of Arizona (fig. 2.4). Alluvial fans and valley floor deposits within broad basins such as the Sulphur Springs Valley in this part of the state (and the rest of the Basin and Range Physiographic Province) typically vary in age from earliest Pleistocene (~2 million years) to late Holocene (less than a few thousand years). Although gravelly alluvium is often the common parent material, soils differ considerably in the development of textural horizons as a function of soil age (Gile 1975a; Gile et al. 1981; Peterson 1981; McAuliffe 1994, 1995, 1999a,b). Extremely young deposits on coarse parent materials may exhibit little or no profile development and often consist of deep, little-weathered sandy loams and loamy sands. With increasing age and clay accumulation, subsurface argillic horizons form. The variation in clay accumulation within argillic horizons yields hydrologic environments ranging from relatively even, deep infiltration and storage of moisture on coarse-textured Holocene soils to relatively shallow infiltration and storage on Pleistocene-aged soils with high clay contents in subsurface argillic horizons. Some younger, Holocene deposits, especially those on valley floors, may be fine-textured because of the silty-clayey nature of alluvium. On some extremely old geomorphic surfaces, partial or complete removal of argillic horizons that were at one time present yields truncated soils, often with considerably coarser texture (Gile 1975b; McAuliffe 1999a).

Soils formed in the Quaternary sediments that fill these basins typically have more horizontally uniform soil hydrologic environments than the fractured and deeply weathered bedrock terrain in which chaparral predominates (e.g., fig. 2.1A and B versus fig. 2.1C). These more horizontally uniform distributions of soil moisture stored nearer the surface are capable of supporting vegetation dominated by perennial C_4 grasses. In southeastern Arizona, grasslands also occupy soils formed on bedrock parent materials, especially on extrusive igneous rocks that form relatively stable slope environments, thereby promoting substantial soil development.

Various species of perennial C_4 grasses differ to some extent in their depths of rooting and soil hydrologic requirements. Consequently, textural characteristics that affect surface infiltration and subsurface percolation of water greatly affect the grass species composition of semidesert grasslands.

The root distributions of various grass species that predominate in soils of different textures reflect the contrasting depth distributions of soil moisture available during the warm growing season (fig. 2.5). For example, clayey soils with slow surface and subsurface infiltration are typically dominated by tobosa and curly mesquite grass. In contrast, black grama *(Bouteloua eriopoda)* often predominates on deep, sandy loams where water percolates much deeper. Such soils commonly occur on young alluvial surfaces as well as on truncated soils of extremely old geomorphic surfaces from which argillic horizons have been stripped, exposing underlying coarse-textured materials. In southeastern Arizona, soils with highly permeable surface horizons and moderate clay accumulation in underlying B horizons often support a wide variety of midheight grasses, including sideoats grama *(Bouteloua curtipendula),* plains lovegrass *(Eragrostis intermedia),* tanglehead *(Heteropogon contortus),* green sprangletop *(Leptochloa dubia),* and others. In the edaphically diverse environments of the American Southwest, the vegetation composition of semiarid grasslands can change markedly over short distances because of subtle soil textural changes (McAuliffe 1995; McAuliffe and Burgess 1995; Burgess 1995).

Environmental Perturbation and Vegetation Change

Soil characteristics not only control vegetation composition, they also strongly influence the kinds of vegetation changes that may occur in response to environmental perturbations. One of the largest historical perturbations that has occurred in semidesert grasslands of the American Southwest has been the heavy use of these environments for grazing by domestic livestock during the past 100–150 years. Considerable increases of woody plants such as mesquite *(Prosopis* spp.) and creosotebush *(Larrea tridentata)* and decrease of cover by perennial grasses have been coincident with this manner of landscape use (Buffington and Herbel 1965; Grover and Musick 1990; Schlesinger et al. 1990; McAuliffe 1998). The causes of some of these changes are complex and involve varying combinations of factors, including competition (or lack thereof) between grasses and woody species, seed dispersal, fire, and changes in soil hydrologic regimes (Burgess 1995).

The reduction of the competitive ability of perennial grasses through overuse of forage resources (Glendening and Paulsen 1955) and the dissemination of seeds by livestock (e.g., mesquite; Brown and Archer 1989, 1999) are principal contributors to these undesirable changes. Yet, the kind

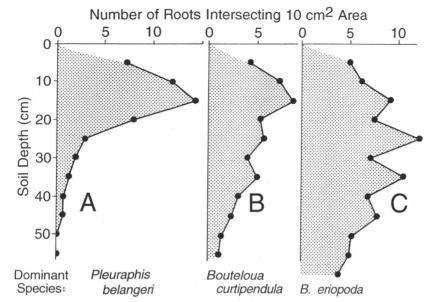

Figure 2.5. Root densities of three grasses in three separate soils located within a radius of 100 m in semidesert grassland at 1,390 m elevation on the alluvial piedmont flanking the northeast side of the Santa Catalina Mountains, Arizona. Each data point consists of the count of the number of roots intersecting three separate 2 cm by 5 cm sample areas on the vertical walls of freshly excavated soil pits. Site A contains a soil with a strongly developed argillic horizon (clay textural class) on an early Pleistocene alluvial fan remnant. Site B is a gentle slope with a less well developed argillic horizon (sandy clay loam textural class). Site C is a truncated soil lacking an argillic horizon and has a B horizon in the loam textural class. Modified from McAuliffe and Burgess (1995).

and magnitude of vegetation changes that have occurred in response to live-stock use vary considerably according to soil conditions. Increases of mesquite and creosotebush, for example, have been far more prevalent on soils that foster deep infiltration, where strongly developed argillic horizons are absent. For example, on the Santa Rita Experimental Range south of Tucson, Arizona, middle-Pleistocene alluvial fan surfaces mantled by soils with thick argillic horizons (maximum clay contents of 31–44 percent) have remained essentially free of woody plants. In contrast, on Holocene surfaces where soils lack significant clay accumulation (maximum clay contents of 4–7 percent), mesquite and other deep-rooted woody plants such as burroweed (*Isocoma tenuisecta*) have markedly increased over the last century (McAuliffe 1995).

A similar relationship between the magnitude of increase of woody species and soil textural characteristics is evident at the Jornada Experimental Range near Las Cruces, New Mexico. The greatest increases in woody plants (mesquite and creosotebush) over the last century have occurred on relatively coarse-textured soils (sands, loamy sands) that are more effective at transmitting the precipitation signal to relatively great depth. Increases of woody plants in finer-textured soils of basin floors occupied by tobosa have been far less prevalent (Reynolds et al. 1999). Even in more mesic regions, soil texture and its impact on the transmission of the precipitation signal deep into the soil affects the propensity for establishment and growth of woody plants in grasslands. Archer (1995) documented the substantial increase of woody plants in southern Texas (mediated by livestock dispersal of seeds) in coarse-textured soils, but such increases did not occur on clayey soils.

The reason for these different propensities for increase of extensive-rooted woody plants in different kinds of soils is probably related to limitations imposed during critical seedling and juvenile stages when root systems are not sufficiently developed to tap into deeper, more constantly available soil moisture. In semiarid environments, the first summer season is a critical time during which a woody plant seedling must secure an uninterrupted supply of soil moisture (Weltzin and McPherson 2000). In coarse-textured soils where moisture from sporadic summer rains reaches slightly deeper layers than in clayey soils, root systems of young plants capable of following and tapping this slightly deeper, persistent moisture may be more likely to survive than those in clayey soils where moisture penetration is always extremely limited and plants are subsequently subject to drought-imposed mortality.

Some of the most serious cases of undesirable vegetation changes have been caused not by the direct impacts of livestock on vegetation, but rather by physical landscape changes, including erosion of soil by water and wind. Soil erosion set into motion by the removal of perennial grass cover varies according to the soils and the parent materials in which these soils form. Some geological environments and their associated soils are very resistant to erosion whereas others are extremely sensitive (Bull 1991; McAuliffe 1995). Extreme alterations of soil hydrologic regimes by soil erosion are usually serious, because their effects are typically irreversible within a time frame relevant to human land use and tenure (McAuliffe 1995).

Predicting the Future

Future climate change is another environmental perturbation whose effects on vegetation will strongly depend on soil characteristics. Any realistic simulation of vegetation changes in response to climate change must consider these variable, soil-dependent responses.

Considerable advances were made during the 1990s in the development of general circulation models (GCMS), Dynamic Global Vegetation Models (DGVMS), and linkages between the two needed for modeling feedback between terrestrial vegetation and climate systems (Neilson and Drapek 1998). These models generally provide geographic resolution at the scale of a 0.5° latitudinal and longitudinal grid (~2500–2700 km²) or larger (Mellilo et al. 1993; Martin 1993; Neilson 1995; Kittel et al. 1995; Pitelka et al. 1997). However, DGVMS have most often treated soils as a computational constant, ignoring the diverse ways in which differences in texture and horizonation control the amounts and depths of infiltration, which in turn can significantly influence vegetation responses. For example, simulation modeling using coarse-scale soils information from the U.S. Department of Agriculture (USDA) 1:7,500,000 soils map for the continental United States yielded considerably different results regarding the predicted distribution of current potential vegetation than did a simulation where soils were assumed to be a ubiquitous sandy loam (Bachelet et al. 1998). This result underscores the necessity of including as accurate and detailed information as possible about soils in any attempt to model vegetation responses to climate change. Failure to do so turns the results of even the most sophisticated, process-based, simulation modeling effort into little more than an extremely complex, computational return to Clements' (1916) concept of the climatic vegetation climax.

Although DGVMS certainly have provided valuable contributions to understanding the possible extent of future ecological changes (e.g., Bachelet et al. 1998; Iverson and Prasad 1998), they probably do not provide the solution to all forecasting and management needs. The most detailed spatial resolution (0.5° grid) of current DGVMS is probably more than adequate to provide information about ecological feedback to the atmospheric climate system. In many cases, however, this scale of spatial resolution is far too coarse to be a useful tool for predicting and managing change by those directly connected with land and resource management. Ranchers, national park superintendents and other managers of public lands, and various other professionals from whom management advice and expertise is sought usually have to make land-use decisions at considerably finer geographic scales. In addition, a principal weakness of most DGVMS is that they predict changes in *potential* natural vegetation rather than *actual* or *possible* vegetation (Martin 1993), thereby disregarding the kinds of ecological changes that have already been wrought by the nearly ubiquitous human inhabitation and use of the planet. It is likely that the greatest ecological changes of the future will not be caused by climate changes alone, but rather by complex synergies between ongoing impacts of land use and climate change (Vitousek 1994). With so much debate continuing in both academic and management communities about the manner in which various human uses alone have impacted ecological systems in arid and semiarid environments, much needs to be done before we can begin to realistically understand and predict these complex interactions between climate change and other perturbations.

A better understanding of vegetation responses to landscape complexity and soils, especially in semiarid environments of the American West, can provide a valuable tool for understanding the conditions, including climate change and land use, under which different kinds of change in actual vegetation can be expected. Such an approach is in some ways analogous to the effort to better understand and predict the types and causes of vegetation changes on U.S. rangelands though the use of state-and-transition models (Westoby et al. 1989; Laycock 1991; USDA 1997). This state-and-transition approach, if applied to efforts to predict the combined influences of climate change and land use on vegetation, presents considerable challenges and requires the expertise and knowledge of many disciplines, including geomorphology, soils, hydrology, ecology, and ecophysiology. Possible ways in which these areas of expertise can contribute are discussed below.

Geomorphology and Soils
Knowledge of the actual physical patterning of *real* landscapes in terms of substrate characteristics is essential for understanding vegetation distributions and potential responses. The diverse edaphic conditions found within regions like the American Southwest are not a random, indecipherable array, but are instead quite predictable given different characteristics of landform morphology, soil parent materials, and landform age (Peterson 1981). The disciplines of Quaternary geomorphology and soil geomorphology have provided a template onto which soil variability (and many ecological responses) can be mapped (Gile et al. 1981; Birkeland 1984). Modern soil surveys and mapping by the USDA Natural Resources Conservation Service have incorporated and used the fundamental knowledge provided by Quaternary geomorphology in predicting and mapping soil occurrences. Furthermore, a detailed map of

Quaternary alluvial deposits within basins often provides the best predictor of vegetation occurrences (McAuliffe 1994, 1995, 1999b).

Soil Hydrology

Better knowledge of the movement and distribution (spatial and temporal) of moisture in soils is essential for predicting plant responses. In nonagricultural soils of arid and semiarid regions, actual measurements of seasonal water dynamics are exceedingly few and restricted to soils formed in relatively fine-grained parent materials (e.g., loess or fine alluvium), where the excavation required for subsoil instrumentation is facilitated. Even in these kinds of edaphic environments, we have little information relevant to the problem of predicting the effects of hydrologic regimes on plant function. The tremendous effort required to empirically monitor soil moisture makes it necessary to develop tools for modeling soil water dynamics.

Process-based soil hydrologic models provide a way to computationally derive water movement, and depth and duration of storage, from detailed (but far more readily acquired) input information on soil properties (horizon thicknesses, texture, bulk density, percent coarse particles, etc.), slope, vegetation (which affects both runoff and evapotranspiration), and the climatic inputs of precipitation and temperature. Several such process-based models have been developed and used to a limited extent in nonagricultural settings in semiarid environments (Flerchinger and Saxton 1989a,b; McDonald et al. 1996; Kemp et al. 1997). With appropriate parameterization and validation of these models for various semiarid and arid environments, these kinds of models could become valuable tools to allow complex (and difficult and costly to directly measure) soil water dynamics to be predicted in diverse soil settings over wide areas. These models also permit a much finer resolution of soil properties and associated hydrologic behaviors than the typical soil hydrologic model components employed in current DGVMs.

A great challenge remains for the understanding of soil water dynamics in extremely rocky terrains. These kinds of substrates underlie a substantial part of nonagricultural lands in the American West. Because of the extremely heterogeneous horizontal and vertical opportunities for water to move through these soils, efforts to model the hydrology of such systems would be extremely difficult, at best. Nevertheless, it may be possible to categorize such complex soil environments within a number of well-defined, discrete categories of different lithologies, depth and type of weathering, and kinds and densities of channels for the conduit of water (fractures, joints, etc.), all of which foster different depth and horizontal distributions of water movement. The statistical association of such multivariate categorical classifications with existing vegetation would greatly improve our knowledge of the combinations of substrate characteristics and climate regimes required to support various kinds of vegetation.

Plant Ecology and Ecophysiology

Any kind of vegetation change requires the establishment of new plants and the death of others. Ecological studies of factors limiting plant recruitment and survival are needed to predict vegetation changes in response to climate change. Investigating seedling and juvenile plant survival in multifactorial field experiments with precipitation input (water augmentation or rainout shelters) and different competitive environments as manipulated variables in different kinds of soils can provide valuable information on the complex interactions that could affect establishment of plants (e.g., Weltzin and McPherson 2000). These kinds of experimental studies need to be combined with detailed studies of soil water dynamics within the experimental areas (empirically measured and modeled). Linking the results of experimental ecological investigations with measured or modeled soil hydrologic dynamics can potentially extend them to a much wider geographic region, with soil hydrologic behavior serving as the common environmental denominator.

Conclusions

By controlling the spatial and temporal distribution of soil moisture, soils exert a great influence on vegetation composition and structure in arid and semiarid lands. Future vegetation responses to climate change will also depend on the nature of underlying soils. However, much work needs to be done to understand linkages between soil hydrologic regimes and plant responses to extend our qualitative understanding of phenomena to a quantitative predictive ability. Realistic, useful predictions of future impacts of climate change should consider these potential impacts within the context of various other pervasive environmental perturbations due to human occupation and use that have greatly affected these environments and will undoubtedly continue to do so. Process-based DGVMs may not provide predictions at the spatial resolution necessary for many on-the-ground management concerns. Many land management needs may be better served with a body of knowledge and predictions that identify thresholds in terms of climatic factors, soil charac-teristics, and land use intensity at which pronounced changes in plant re-sponses are expected (e.g., success vs. failure in seedling establishment). The gathering, synthesis, and organization of this kind of informational frame-work presents tremendous challenges, but could prove extremely useful for predicting future changes in arid and semiarid lands of the world. However, it must be remembered that even with the best information available and considerable effort, it is very possible, perhaps even likely, that future climate changes will produce completely unpredictable surprises (Overpeck 1996). Broad, multidisciplinary approaches involving collaborative bridges between the atmospheric, earth, and ecological sciences perhaps offer the most promis-ing ways to discover some of these potential surprises.

Acknowledgments
Carla McAuliffe created elevational cross-sections used in the preparation of figure 2.4 with digital elevation models [DEM Lab, 2000, Arizona Board of Regents] which are produced from USGS 1:250,000 scale digital elevation mod-els. She also edited the original manuscript. I thank J. Weltzin, G. McPherson, and an anonymous reviewer for their helpful comments and suggestions on the manuscript.

3

Responses of Woody Plants to Heterogeneity of Soil Water in Arid and Semiarid Environments

DAVID G. WILLIAMS & KEIRITH A. SNYDER

Woody plants are conspicuous and functionally important components in arid and semiarid ecosystems. Vegetation responses at the scale of individual plants (e.g., growth, gas exchange, resource allocation) provide an important linkage between environmental change and ecosystem dynamics. The water relations, rooting patterns, and environmental tolerances of woody species are key elements in hydrologic and dynamic vegetation models used to predict climate-vegetation interactions at local, regional, and global scales (Stephenson 1990; Sellers et al. 1997; Daly et al. 2000; Jackson et al. 2000; Reynolds et al. 2000).

Many (but not all) woody species in arid and semiarid environments are capable of extracting water from a large soil volume and from great depths. These "extensive exploiters" (see McAuliffe, chapter 2, this volume) often develop a dimorphic rooting pattern to capture water in shallow soil layers during rainy periods and from deeper soil layers during periods of drought. Taproots of mesquite *(Prosopis)* were uncovered 50 m below the surface in an open mine pit in Arizona (Phillips 1963), and roots of woody species in the Kalahari Desert have been observed to 63 m (Jennings 1974). Such extensive root systems act to buffer the plant from short-term deficits in precipitation, but are costly in terms of the energy required for their construction and maintenance.

Conversely, grasses and other herbaceous species, the "intensive exploiters," develop very shallow and dense root systems that generally extend no deeper than about 0.5 m (Jackson et al. 1996). Photosynthetic gas exchange and growth of shallow-rooted herbaceous species are restricted to rainy periods, when shallow soil layers contain adequate levels of available water and mineral nutrients. Seasonal patterns of growth and gas exchange of deeply rooted shrubs and trees do not necessarily coincide with the rainy season, because of the persistence of available moisture deep in the soil. Consequently, the presence of deeply rooted woody species in a community can extend the period of production and transpiration beyond rainy periods, allowing these species to become the dominant cover in many arid and semiarid ecosystems.

This very simplified view of the architecture and function of woody plant root systems is instructive, but ignores the tremendous level of intra- and interspecific variation for active rooting depth often observed in this life form. As stated above, belowground processes involving woody plant root systems are key elements in models of vegetation change and ecosystem function and dynamics (Daly et al. 2000; Jackson et al. 1996, 2000). Our ability to scale processes from individual plants, species, and sites across heterogeneous desert landscapes will depend to some degree on how intra- and interspecific variation is treated in these models.

Pulses of moisture in arid and semiarid environments are distributed heterogeneously in time and space. Patterns of precipitation, rates and patterns of evapotranspiration, structure and composition of the vegetation, local topography and drainage patterns, and the complexity of the soil substrate influence the duration and spatial distribution of pulses of soil moisture (see chapter 2). Global changes that alter the seasonal timing and distribution of available soil moisture can have significant impacts on woody plants and the ecosystem processes that these plants control. This chapter addresses the significance of intra- and interspecific variation in water use by woody plants in arid and semiarid ecosystems. We address the following questions: How much do species and populations of woody plants differ in their ability to take advantage of precipitation events during the growing season and capture moisture within a heterogeneous soil environment? What mechanisms account for this variation? Can we expect fairly uniform responses by species and populations to precipitation change across different habitats or under future conditions of elevated atmospheric carbon dioxide (CO_2) concentration? Answering these questions requires a thorough examination of how woody plants exploit pulses of water that rapidly emerge and disappear within the soil environment. We use examples mostly from the American Southwest, although our synthesis applies to all semiarid and arid ecosystems.

Heterogeneity of Soil Water in Arid and Semiarid Environments

Arid and semiarid ecosystems are characterized by low resource levels, with water representing the most limiting of these resources for plant growth and ecosystem productivity (Noy-Meir 1979). Water in these dry environments is present in the soil as a series of pulses that vary in timing and duration depending on many biotic and abiotic factors. An overriding abiotic control on the spatial and temporal distribution of soil water in these environments is climate. The seasonal pattern of precipitation and evapotranspiration largely determines the temporal dynamics and vertical heterogeneity of soil water within the rooting zone of woody species. Regions with bimodal precipitation, with rain falling in summer and winter, have complex patterns of soil moisture dynamics. Winter precipitation from slow-moving frontal systems falls over extended periods and penetrates to deep soil layers in part accentuated by low evaporative conditions. Conversely, convective storms during summer periods generally do not wet deep soil layers. High rates of evaporation and short, intense rainfall events tend to restrict summer-moisture pulses to upper soil layers. Consequently, moisture pulses of high amplitude and short duration occur in upper soil layers during the warm growing season, whereas low-amplitude pulses of longer (seasonal) duration from winter inputs characterize soil layers below 50 cm (Scott et al. 2000; Kemp et al. 1997).

At a more local level, landscape topography and soil structural and textural characteristics have important impacts on the spatial and temporal distribution of soil water (see chapter 2). Combined with these substrate controls, vegetation affects water input and loss from different soil layers. Woody plants affect the temporal and spatial heterogeneity of soil water in arid and semiarid ecosystems through interception, uptake by roots, hydraulic redistribution by the root system, and alteration of microclimate. Soil water is removed by plants during daytime transpiration, but plant canopies and roots redistribute precipitation, runoff, and soil moisture. For instance, water may be transferred nocturnally between different locations in the soil through roots by "hydraulic lift" (Caldwell et al. 1998). More accurately termed "hydraulic redistribution," this process can transfer soil moisture from deep to shallow layers and from shallow to deep layers depending on the direction of soil water potential gradients and the spatial distribution, density, and permeability of fine roots (Richards and Caldwell 1987; Dawson 1993; Caldwell et al. 1998; Burgess et al. 1998, 2001; Hultine et al. submitted). This mode of unsaturated flow is not accounted for in hydrologic models, but may have significant impacts on landscape hydrologic balance (Dawson 1993).

Horizontal distribution of soil moisture is strongly affected by the presence of woody species. Interception losses and extraction of soil water by roots tend to reduce soil moisture beneath woody plant canopies (Breshears, Rich, et al. 1997; Breshears et al. 1998), whereas runoff redistribution and localized shading enhance soil moisture in these locations (Joffre and Rambal 1988). These opposing processes tend to accentuate temporal variation of soil moisture across inter- and intracanopy locations.

Together, precipitation patterns, soil characteristics, topography, and the activities of plants themselves create a heterogeneous and dynamic belowground resource environment for woody plants in arid and semiarid ecosystems. Patches of soil water emerge and disappear at different rates and in different locations in the soil. The structure and physiology of woody plant root systems and the flexibility of root-system response govern how effectively woody plants exploit these transient pulses of water, which in turn determine the magnitude and pattern of plant-plant and plant-environment interactions and the dynamics of plant communities and ecosystems (Schwinning and Ehleringer 2001).

Water Movement into and through Roots of Woody Plants

Plant roots sense gradients of soil water potential and grow toward patches of moist soil. Elongation of the dry side of a growing root is stimulated, causing the root to grow toward the wet patch (Takahashi 1994). Once roots occupy a region or layer of the soil, the absorptive area, the root hydraulic conductivity, and the water potential gradient from soil to and through the roots will determine how rapidly a pulse of moisture is taken up by the plant. However, water potential gradients and resistances to flow change rapidly as the soil dries. Soil pore spaces fill with air causing restrictions in flow to the root, air gaps develop at the root surface as roots shrink, and conductivity of the root itself drops in drying soil (Passioura 1988; Nobel and Cui 1992a,b; Nobel 1994; Steudle and Peterson 1998; Hacke et al. 2000).

Axial flow of water through roots occurs almost entirely in the xylem. Most trees and shrubs in arid and semiarid regions are either conifers or dicots and thus can add new xylem to roots through secondary growth. Water flux through these roots, therefore, should not be limited by low axial conductance (Passioura 1988). However, observed reductions in axial conductivity of roots may explain the sharply reduced hydraulic conductance of the soil-plant pathway in drying soils. Roots often are more vulnerable than shoots to xylem cavitation induced by drought, and the water potentials necessary to induce root cavitation can differ markedly among woody species (Alder et al. 1996; Mencuccini and Comstock 1997; Linton et al. 1998). For example, surface roots of Colorado pinyon *(Pinus edulis)* are more vulnerable to drought-induced xylem cavitation than are roots of co-occurring Utah juniper *(Juniperus osteosperma)* (Linton et al. 1998).

There may be some advantage to partial root cavitation and reduction of root hydraulic conductance as soil dries. Partial root cavitation may regulate the rate of water extraction from the soil such that hydraulic failure in the soil-to-root pathway is avoided (Alder et al. 1996; Sperry et al. 1998; Hacke et al. 2000). If water is extracted too rapidly from a drying region in the soil, localized reductions in soil water content and hydraulic conductance may limit plant transpiration.

Hydraulic characteristics of the entire root system may further determine patterns of water extraction from different soil layers or patches. For instance, the hydraulic conductance of deep taproots and shallow lateral roots in the shrub snakeweed *(Gutierrezia sarothrae)* can differ markedly (Wan et al. 1994). Hydraulic conductivity of taproots in this species is more than threefold higher than in shallow lateral roots during the dry season. The suberized lateral roots possess fewer large xylem vessels than do tap roots, and after rewetting, lateral roots do not regain hydraulic conductances that match those of the deeper taproots. Consequently, water is taken up preferentially from deeper roots than from shallow roots in this shrub.

Spatial and Temporal Heterogeneity of Water Uptake by Woody Plants

There are several ways to obtain information on the spatial and temporal patterns of water uptake from woody plant roots. Traditional methods involve root trenching in combination with sequential measurements of volumetric soil water content. Differences in depletion of soil water around trenched and nontrenched plants is used to quantify root uptake from different soil layers. This method is laborious and expensive and only provides indirect information on plant water sources. One alternative approach relies on tracers to follow the path of water movement from different soil locations to the plant. The most suitable tracer for water movement in the soil-plant-atmosphere system is water itself. Stable isotope ratios (^{2}H/^{1}H and ^{18}O/^{16}O) of water are now widely measured for this purpose (Ehleringer et al. 1991; Evans and Ehleringer 1994; Lin et al. 1996; Kolb et al. 1997; Williams and Ehleringer 2000; Snyder and Williams 2000). Although it is relatively straightforward to label a region of the soil with enriched levels of deuterium (^{2}H or D) or ^{18}O and detect its movement through plants, this is not always necessary. The stable isotope ratio of water in different soil layers often varies naturally because of (1) enrichment of heavy isotopes during evaporation from the surface layers (Allison et al. 1983; Barnes and Allison 1983, 1988) and (2) seasonal and storm-related isotopic variation in precipitation inputs (Gat 1980; Ingraham and Taylor 1991). Soil water originating from summer and winter precipitation can differ substantially in ratios of hydrogen and oxygen isotopes (Dansgaard 1964). Winter precipitation that percolates to deep soil layers often is depleted in heavy isotopes (^{2}H and ^{18}O), whereas summer precipitation that wets only the upper soil layers is subject to evaporative processes that leave it enriched in these isotopes (Allison et al. 1983; Ingraham et al. 1991).

Except in some halophytes (Lin and Sternberg 1993), the isotopic composition of water taken up by plant roots and transported to shoots remains unaltered until it reaches sites of evaporation, such as leaf surfaces (Dawson and Ehleringer 1993; Brunel et al. 1995). Thus, for most woody species, measurement of stable isotope ratios at natural abundance levels provides a simple and relatively inexpensive way to assess water sources (Ehleringer and Dawson 1992; Brunel et al. 1995).

It is often assumed that woody species in arid and semiarid environments extract water from deep soil layers and thus interact only minimally with shallowly rooted herbaceous species (Walter 1971). It is further assumed that most plants will absorb water from upper soil layers when these layers are wet. Many exceptions to these generalizations have been found. Stable isotope studies have shown that woody species respond very differently to the magnitude and seasonal input of moisture pulses (Ehleringer et al. 1991; Flanagan et al. 1992; Valentini et al. 1992; Lin et al. 1996; Weltzin and McPherson 1997; Williams and Ehleringer 2000). Annuals and herbaceous perennials in a cold desertscrub community in southern Utah tended to take up summer precipitation from shallow soil layers, whereas most woody perennials during the hot summer took up deeper soil water (Ehleringer et al. 1991). However, some woody species in this community used summer precipitation and had hydrogen isotope ratios that were identical to those of several of the herbaceous perennials. Other co-occurring woody species in this desertscrub community utilized only small fractions of summer rain; surface roots in these species apparently were not active during the hot summer monsoon period.

Riparian trees in arid and semiarid regions are sustained by shallow water tables. However, differences in the ability to use pulses of summer precipitation are found even among these phreatophytes. Smith et al. (1991) observed temporal variation for use of shallow soil water by riparian trees growing along a perennial montane stream in California; trees used shallow soil water in spring and then shifted to groundwater during drier summer months. In western Arizona, Frémont cottonwood *(Populus fremontii)* and Goodding willow *(Salix gooddingii)* used groundwater throughout the entire growing season, but *Tamarix ramoisissima,* an invasive species, used shallow soil water late in the growing season (Busch et al. 1992). In southeastern Arizona, the proportion of transpiration water derived from summer rainfall varied widely among six dominant woody riparian species growing along an intermittent stream (fig. 3.1) (Snyder et al. 1998). Facultative phreatophytes such as netleaf hackberry *(Celtis reticulata),* velvet ash *(Fraxinus velutina),* and velvet mesquite *(Prosopis velutina)* used a greater proportion of shallow soil water than did the obligate phreatophytes Frémont cottonwood, Goodding willow, and seep willow *(Baccharis glutinosa).* Even among obligate phreato-phytes, substantial variation in the use of summer precipitation is observed. For example, Frémont cottonwood consistently used a greater proportion of shallow soil water than Goodding willow on the San Pedro River in south-eastern Arizona during the summer rainy season (Snyder and Williams 2000). Therefore, even in riparian environments where woody species are primarily dependent on groundwater, precipitation pulses supply substantial amounts of transpirational water to some species under some conditions.

Substantial interspecific variation in use of summer precipitation pulses is observed in trees of upland environments in semiarid landscapes. For example, woody species in pinyon-juniper-oak communities in the south-western United States exhibit considerable variation in active rooting depth. In the pinyon-juniper community type, there are deep and shallowly rooted shrubs and trees, as well as species that take up water from throughout the soil profile. Utah juniper *(Juniperus osteosperma)* and Colorado pinyon *(Pinus edulis)* derive most of their water from shallow soil layers (< 30 cm) during the summer rainy periods, whereas co-occurring woody dicots such as big sagebrush *(Artemisia tridentata),* rabbit-brush *(Chrysothamnus nauseosus),* and Gambel's oak *(Quercus gambelii)* take up water from deep soil layers, but use

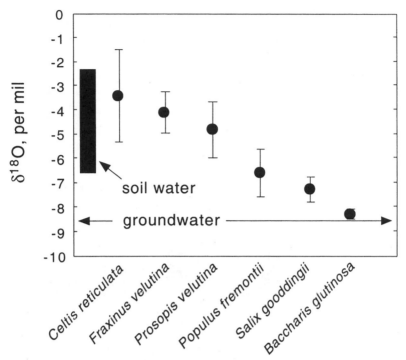

Figure 3.1. Oxygen stable isotope ratios ($\delta^{18}O$) of tree xylem water, groundwater, and soil water in a riparian ecosystem near Tucson, Arizona. Plant species are netleaf hackberry *(Celtis reticulata)*, velvet ash *(Fraxinus velutina)*, velvet mesquite *(Prosopis velutina)*, Frémont cottonwood *(Populus fremontii)*, Goodding willow *(Salix gooddingii)*, and seep willow *(Baccharis glutinosa)*. Data are means and standard error bars (n = 3–5 trees). Data were collected in August 1996 following a large precipitation event. From Snyder et al. (1998). Used with permission.

very little summer precipitation (Flanagan et al. 1992; Evans and Ehleringer 1994; Williams and Ehleringer 2000).

The size of precipitation events during the growing season is an important factor that differentiates pulse utilization by woody species in the pinyon-juniper-oak community. Williams and Ehleringer (2000) performed an irrigation experiment with deuterium-labeled water at a site on the high Coconino Plateau near the south rim of the Grand Canyon to characterize the sensitivity of surface roots of Colorado pinyon, Utah juniper, and Gambel's oak to artificial 10 mm and 25 mm summer precipitation events. Gambel's oak did not respond to either 10 mm or 25 mm irrigations (fig. 3.2). Although Gambel's oak develops a network of surface roots (Clary and Tiedemann 1986), these roots apparently do not take up significant amounts of water

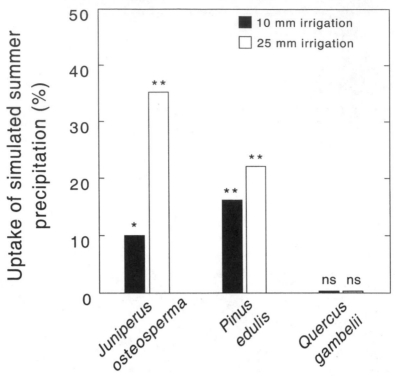

Figure 3.2. Percentage of tree transpiration derived from 10 mm and 25 mm simulated summer rain events in a pinyon-juniper ecosystem near the south rim of the Grand Canyon, Arizona. Percentages were calculated using hydrogen stable isotope ratios (δD) of tree xylem water. Xylem water was collected five days after trees had been irrigated with deuterium-labeled water. Each bar represents the mean value from four to five plants after adjusting for natural changes in D of control plants over the five-day experiment. Statistical significance values (*$P < 0.05$, **$P < 0.01$) are from t tests between means of irrigated and nonirrigated plants. From Williams and Ehleringer (2000). Used with permission.

during the summer growing period, at least at this site. Colorado pinyon was the only species to respond significantly to the very small 10 mm irrigation, but response to the 25 mm irrigation was greatest in Utah juniper.

Vertical patterns of water uptake by woody species in arid and semiarid environments also can be strongly controlled by plant developmental stage (Dawson and Ehleringer 1991; Donovan and Ehleringer 1992). The active rooting depth of seedlings and young saplings of woody plants is often restricted to shallow soil layers. For example, 1- and 2-year-old *Quercus emoryi* seedlings within a temperate savanna ecosystem in southern Arizona obtained water only from shallow soil layers, which suggests that competition for soil water between these young trees and neighboring grasses occurs for at least two years after germination of the oak (Weltzin and McPherson 1997). Furthermore, 2-month-old seedlings of this oak used water from shallower depths in the soil profile than grasses, which may facilitate its survival and early establishment. Larger adult *Q. emoryi* plants used water from well below the active rooting depth of the grasses. Woody species in arid and semiarid environments must cope with extreme fluctuations in shallow soil moisture during the seedling and sapling developmental stages. Therefore, predictions about woody plant response to precipitation change must consider the resource environment of seedlings as well as that of adults.

Most studies conducted in arid and semiarid ecosystems have focused on responses of woody species to variation of soil water in vertical space or through time, but these species may also differ in active rooting distribution in horizontal space. For example, horizontal patterns of root distribution and soil water uptake were found to differ between Colorado pinyon and one-seed juniper *(Juniperus monosperma)* in New Mexico (Breshears, Myers, et al. 1997). One-seed juniper responded more to localized surface additions of water to intercanopy areas than did pinyon. Such horizontal variability in water uptake could shape patterns and magnitudes of species interactions and should be incorporated into models of community and ecosystem change (Breshears, Myers, et al. 1997; Breshears and Barnes 1999).

Water Uptake within a Heterogeneous Soil Environment

Water content fluctuates dramatically in the top 30 cm of the soil during summer rainy periods. To use this moisture, woody plants must either maintain absorptive roots in shallow soil layers or proliferate new roots into these layers when moisture arrives after rain events. In either case, water uptake from a patch of moist soil comes at a physiological cost to the plant. Using economic analogy, Bloom et al. (1985) predicted that plants will continue to produce roots to acquire a limiting soil resource as long as the cost of root production is exceeded by the benefit of water uptake from those roots. Costs and benefits are best measured using the limiting resource as the common "currency." In the case of roots growing into a patch or layer of moist soil, the water cost (water expended to produce the roots) is compared to the water income (water uptake by the roots). Absorptive capacity of roots is often inversely related to root longevity and construction cost (Nobel 1994). As such, plants can either produce long-lived roots with relatively low absorptive capacity or short-lived roots with high absorptive capacity. Either strategy may be favored depending on environmental conditions and the dynamics of moisture pulses. Plants often produce both kinds of roots and thus may optimize water uptake through changes in the proportion and timing of production of coarse, long-lived roots and fine, short-lived roots (Eissenstat and Yanai 1997).

Since plants "trade" water for carbon at stomata during photosynthesis, the carbon cost of producing and maintaining roots can be converted to water cost with water-use efficiency (WUE; CO_2 assimilation rate divided by transpiration rate at leaf surfaces). Water cost per unit mass of root produced (C_w, mol H_2O g^{-1} root) is calculated as

$$C_w = (C_c + R_m T)/WUE \quad \text{[eq. 1]}$$

where C_c is carbon construction cost of roots, or the amount of simple carbohydrate (e.g., glucose) required to build the root (Williams et al. 1987); R_m is root maintenance respiration; and T is the time interval over which costs and incomes are measured (modified from Hunt et al. 1987). Root efficiency (E) is then calculated as

$$E = I_w/C_w \quad \text{[eq. 2]}$$

where I_w (= income) is the total amount of water absorbed from the roots (mol H_2O g^{-1} root) over the defined time interval. The time interval over which costs and incomes are assessed is important because I_w and C_w accrue at different rates. Costs accrue initially very rapidly as the roots grow. Thus, production of new roots in a soil layer will accrue net benefits ($I_w/C_w > 1$) only after an indefinite period of time, depending on the moisture availability within the soil layer, root uptake rates, and WUE.

The above model does not account for costs associated with carbon exudation from roots, maintenance of symbiotic relationships, tissue loss to herbivores, the cost of building and maintaining transport roots, or temporal disequilibria associated with resource storage by the plant (Eissenstat and Yanai 1997). Nevertheless, plants should produce roots in a patch of moist soil if water income from the root in the patch is greater than the cost of building and maintaining the root. However, since there is limited carbon available for root growth, plants may grow roots preferentially into soil layers with stable, reliable sources of water (e.g., groundwater), such that root efficiency (E) is maximized. The costs of root construction and maintenance may be very different in different soil layers. Furthermore, roots compete internally for limited plant carbon and allocation is determined by source-sink relationships among different roots (Eissenstat and Yanai 1997). Plants should preferentially allocate carbon to roots in surface layers when root efficiency (E in eq. 2) of shallow roots exceeds that in deeper roots ($E_{shallow} > E_{deep}$). The reverse would be expected in riparian environments where groundwater is available within the rooting zone. Thus, root growth in one region of the soil may come at the expense of root growth elsewhere.

To maximize root efficiency in soil regions that experience wide moisture fluctuations (e.g., surface soil layers), root costs should be minimized when water is not available. Roots of succulent CAM (Crassulacean acid metabolism) species in the Mojave and Sonoran Deserts have high WUE and very low root construction (C_c) and maintenance respiration (R_m) costs (Nobel et al. 1992). These characteristics may allow CAM succulents to persist in very arid environments where pulses of summer precipitation are of very short duration and magnitude.

High root efficiency also may arise through changes in root life span. Long-lived roots should be favored in stable moisture patches since the initial investment (C_c) is a one-time fixed cost. However, it is beneficial to produce short-lived roots or no roots in soil locations characterized by infrequent or short-duration moisture pulses. There may be too little time to recover costs of roots growing in these ephemeral moisture patches. Some degree of plasticity in root life span and cost therefore may also be favored in environments with unpredictable moisture availability. There is limited information on such plasticity for woody species in arid and semiarid ecosystems. However, plasticity in root longevity in woody species is common in other ecosystems; roots produced in response to experimental additions of water lived longer than roots in control treatments in a mixed hardwood forest in Michigan (Pregitzer et al. 1993).

Use of summer precipitation pulses tends to be greater among drought-tolerant, facultative phreatophytes compared to drought-intolerant, obligate phreatophytes in riparian ecosystems (Busch et al. 1992; Williams et al. 1998; Snyder and Williams 2000). We predicted that construction costs of shallow roots would be lower in facultative phreatophytes compared to obligate phreatophytes in desert riparian ecosystems, where surface moisture is very dynamic and of short duration. Low root construction costs would allow facultative phreatophytes to quickly recover root investment in shallow soil layers. Roots of five riparian woody species were collected in 1998 from 10 to

30 cm depth in alluvial soils along an ephemeral drainage in southeastern Arizona. Contrary to expectations, construction costs (C_c) of 5–10 mm diameter roots in the facultative phreatophytes were *higher* than those in roots of the obligate phreatophytes (fig. 3.3; Williams, unpublished data). Root carbon construction cost is but one of several parameters that determine root efficiency (eqs. 1 and 2). Information on root respiration rates, water uptake capacity, and root longevity are needed to fully evaluate root efficiency (E) for woody species in these desert washes. Compared to roots of obligate phreatophytes, roots of facultative phreatophytes may persist longer into a drought and survive repeated episodes of soil drying due to these species lower vulnerability to water-stress induced cavitation (Pockman and Sperry 2000). Although roots of facultative phreatophytes may be energetically expensive to construct, their duration and absorptive capacity at low soil water contents may favor their existence in shallow soil layers compared to the obligate phreatophytes that appear not to use shallow soil water.

Nitrogen concentration in root tissue was strongly correlated with C_c among riparian tree species used in our root construction cost study (fig. 3.3). High nitrogen in the roots of the facultative phreatophytes may be associated with greater allocation of this soil resource to nitrogen-containing defensive compounds or to enzymatic activity associated with mineral nutrient absorption and assimilation. The latter highlights the role of roots in arid and semiarid environments for functions other than water absorption.

Further evidence that woody plants forage optimally for heterogeneously distributed soil moisture comes from studies on intraspecific variation in water source use. Dawson and Ehleringer (1991) found that large trees in a riparian ecosystem in Utah did not take up surface water (stream and precipitation), but instead used only the more stable groundwater. However, juvenile trees with inherently shallower rooting depths used stream water and precipitation. Juvenile boxelder trees *(Acer negundo)* likewise used a greater proportion of surface water than large boxelder trees along a perennial stream in north-central Arizona (Kolb et al. 1997). Similarly, in southeastern Arizona, the riparian tree species cottonwood *(Populus fremontii)* and mesquite *(Prosopis velutina),* but not the willow *(Salix gooddingii),* used proportionally greater amounts of summer precipitation as depth to groundwater increased (fig. 3.4) (Snyder and Williams 2000). Access to a stable groundwater supply apparently causes these desert riparian species to redirect limited plant resources to growth and activity of deep roots by sacrificing activity of shallow roots.

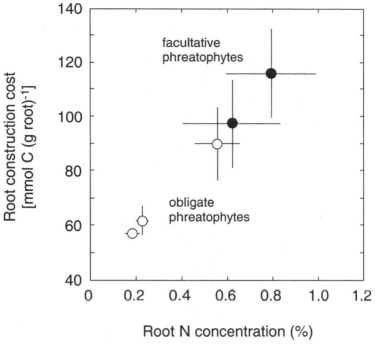

Figure 3.3. Construction cost [mmol C (g root)⁻¹] and nitrogen concentration (percent) of roots (dry mass basis) collected from five riparian tree species growing along an ephemeral drainage in southeastern Arizona. The tree species are from lower left: seep willow *(Baccharis glutinosa)*, Frémont cottonwood *(Populus fremontii)*, Goodding willow *(Salix gooddingii)*, netleaf hackberry *(Celtis reticulata)*, and velvet mesquite *(Prosopis velutina)*. Construction cost was estimated from heat of combustion, ash-free dry mass, and root nitrogen (Williams et al. 1987). Data are means and standard error bars.

The average amount of moisture available over the growing season in different soil layers should determine the optimal distribution and activity of roots through the soil profile among populations within woody species. As the amount of summer precipitation increases, as it does along natural summer precipitation gradients or as it might over time as climate changes, a threshold level of input may be reached at which populations of woody plants will alter their active rooting depth and deploy roots in upper soil layers to capture surface moisture (Ehleringer and Dawson 1992). This prediction was tested in woodland communities in Utah and Arizona across a region that spans a large summer rainfall gradient (Williams and Ehleringer 2000). Populations of Colorado pinyon, Utah juniper, and Gambel's oak from eastern Arizona to northern Utah were compared for summer precipitation use using stable isotope tracers. Fractional uptake of moisture from upper soil layers during

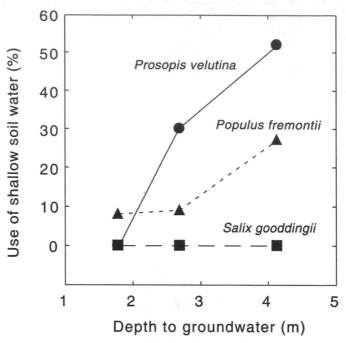

Figure 3.4. The relationship between depth to groundwater and the percentage of tree transpiration derived from shallow soil layers (0 to 50 cm) for Goodding willow *(Salix gooddingii)*, Frémont cottonwood *(Populus fremontii)*, and velvet mesquite *(Prosopis velutina)*. Percentages were calculated using hydrogen isotope values (δD) of tree xylem water. Data were collected at three sites along the San Pedro River in southeastern Arizona that differed in depth to groundwater. From Snyder and Williams (2000). Used with permission.

the summer rainy period increased sharply in a nonlinear fashion for these species as summer rain input increased (fig. 3.5). The threshold level of summer rain needed for activation of surface roots apparently differed among the three species; Gambel's oak required greater inputs of summer rains than did pinyon and juniper before root activity in surface layers was detected using our tracer approach. Variation in these thresholds is expected since root costs and uptake capacities differ among species, and this determines the point along the gradient where $E_{shallow} > E_{deep}$.

Stable isotope tracers only characterize the fraction of transpiration water derived from different moisture sources. In a southeastern Arizona riparian community, Frémont cottonwood takes up summer precipitation from shallow lateral roots, and the fraction of tree transpiration derived from these roots increases across a gradient of declining groundwater levels (fig. 3.4; Snyder and Williams 2000). The amount or rate of water extraction by roots

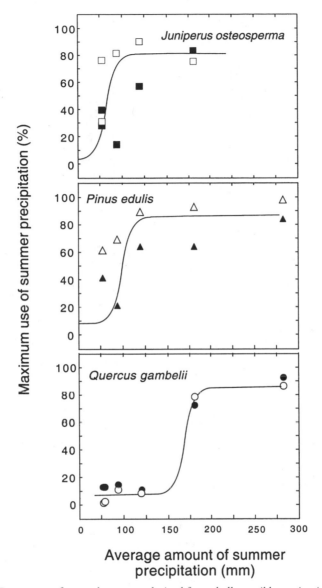

Figure 3.5. Percentage of tree xylem water derived from shallow soil layers in pinyon-juniper communities during the summer monsoon season of 1993 (closed symbols) and 1994 (open symbols). Data are plotted as a function of site-averaged summer precipitation amount. Sites spanned a large gradient in summer precipitation from eastern Arizona to northern Utah. Trees sampled were *Juniperus osteosperma*, *Pinus edulis*, and *Quercus gambelii*. Percentages were calculated using hydrogen isotope values (δD) of tree xylem water. From Williams and Ehleringer (2000). Used with permission.

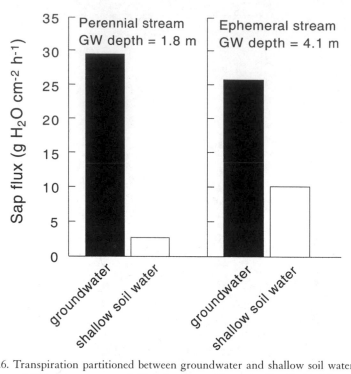

Figure 3.6. Transpiration partitioned between groundwater and shallow soil water for two populations of Frémont cottonwood *(Populus fremontii)* on the San Pedro River, Arizona. Data expressed as transpiration flux per unit sapwood are of the trees (g H_2O cm^{-2} sapwood h^{-1}). Values were calculated from midday sap flux measurements and water source data from isotope measurements. From Snyder and Williams (2000). Used with permission.

from a soil layer is more appropriate for analyzing tradeoffs and root efficiencies. Simultaneous measurements of sap flow (Schaeffer et al. 2000) on the same cottonwood trees that were sampled isotopically for water sources (Snyder and Williams 2000) revealed that this obligate riparian species of desert environments had lower rates of water uptake from shallow soil layers at a site where groundwater was less than 2 m and freely available to the trees than at sites where the water table was greater than 4 m (fig. 3.6). Apparently the activity of shallow roots was sacrificed by cottonwood when deep moisture was easily obtained, as is predicted from the model.

Soil Water Uptake by Woody Species in Future Environments

Global changes resulting in alteration of the amount or timing of precipitation, such as rising atmospheric CO_2, will have direct impacts on water uptake patterns by woody plants in water-limited ecosystems. Carbon dioxide concentrations in the atmosphere are predicted to rise to more than 700 ppm by the end of the twenty-first century, which will directly affect woody plant photosynthesis. Rising CO_2 levels may alter (either directly or indirectly) woody plant hydraulic architecture, the distribution and density of fine roots, leaf area, transpiration, and wue (Tyree and Alexander 1993; Poorter 1993; Field et al. 1995; Polley et al. 1997; Ceulemans et al. 1999; Pregitzer et al. 2000). Enhanced wue under elevated CO_2 will increase root efficiency (E) by lowering the water cost (C_w) of roots (eq. 1). Anatomical changes induced by rising CO_2 may alter root and shoot hydraulic conductivity and vulnerability to cavitation (Tyree and Alexander 1993), further changing E (through changes in I_w; eq. 2) over a drought cycle. As CO_2 increases, plant growth may become less limited by carbon. Therefore, roots in different regions of the soil may compete less for plant-available carbon, resulting in changes in belowground patterns of root allocation.

Reduced stomatal conductance and evapotranspiration under elevated CO_2 may enhance availability of moisture in shallow soil layers (Owensby et al. 1993; Field et al. 1997). Enhanced moisture content in surface layers may cause woody species with flexible allocation responses to shift rooting activity to these layers. Trees grown under elevated CO_2 indeed have been found to initiate more lateral roots to enhance the capture of shallow soil moisture (Ceulemans et al. 1999). However, enhanced moisture in surface layers may increase the potential for water to percolate to deeper layers and favor greater exploration and growth by deep roots (Polley et al. 1997). It is not clear under this scenario if the relative rooting activity at different depths would change.

Conclusions

Soil water is heterogeneously distributed within the rooting zone of woody plants in arid and semiarid ecosystems. Soil water is generally available as a series of pulses that vary greatly in timing, duration, and location within the soil profile. Woody plants integrate this patchiness with their extensive root systems. However, growth of roots into moist soil layers comes at a cost, and plants have finite resources to allocate to capturing transient moisture pulses. There is considerable intra- and interspecific variation in the ability of woody plants to capture pulses of soil water from precipitation that occurs during the growing season. At the landscape scale, habitat characteristics strongly control patterns of root system development and exploitation of moisture pulses. At sites where groundwater is an important water source, its depth and reliability determine the extent of surface root activity and use of growing-season precipitation. Threshold amounts of summer precipitation input apparently are needed for activity of water-absorbing roots in surface layers, and species apparently differ in their sensitivity to these inputs. Finally, the direct impact of rising atmospheric CO_2 could alter belowground resource uptake by woody plants in arid and semiarid ecosystems; root growth, turnover, distribution, and hydraulic conductivity are all potentially affected by this greenhouse gas.

Global changes may include altered temperature, precipitation, CO_2 concentration, and nitrogen inputs in terrestrial ecosystems. Together, these changes would have very complex effects on, and interactions with, water uptake patterns in dominant woody species. There is considerable uncertainty about how to treat belowground processes in hydrologic and global change models. Yet these processes drive vegetation responses to precipitation change. The tremendous variation in woody species responses to spatial and temporal heterogeneity in soil moisture adds another level of complexity to our conceptual understanding of plant responses to environmental change. We often observe as much variation in active rooting distribution within species as across species of woody plants. Our ability to predict belowground responses of woody plant species to precipitation change and the consequences of plant responses for surface hydrologic balance will depend to some degree on how intra- and interspecific variation is treated in ecosystem models.

Acknowledgments
We gratefully acknowledge Will Pockman, Susan Schwinning, and Jake Weltzin for the many insightful and instructive comments on earlier versions of the chapter. We thank Dan Koepke and Kelly Veach for field and lab assistance with root construction cost estimates of riparian trees. Work on root construction costs was supported by a USDA–NRI grant to DGW.

4

The Importance of Precipitation Seasonality in Controlling Vegetation Distribution

RONALD P. NEILSON

The potential for human-induced global warming has stimulated an interest, among both scientific and lay people, of the possible responses of the terrestrial biosphere to changes in global temperature and rainfall patterns. There are well-recognized correlates between vegetation distribution and global patterns of average annual temperature and total annual rainfall distributions. However, the importance of seasonal precipitation patterns is not as well understood. Yet, the seasonal patterns of precipitation are expected to change across different regions of the Earth's surface, due to global warming.

It has long been recognized that climate is the primary driver of the spatial patterns of organismal distribution. The distributions of the major biotic zones, such as tundra, boreal, temperate, subtropical, and tropical, bear clear relation to the large-scale Hadley circulation of the atmosphere, primarily via the atmospheric circulation control on the annual and seasonal patterns of temperature and precipitation. Most early biogeographic research drew correlates between broad vegetation zones and annual average temperature and total annual precipitation. Among the earlier approaches were the well known Köppen (Bailey 1996) and Holdridge (1947) schemes, each of which relied primarily on annual temperature and precipitation patterns. The role of precipitation seasonality, although recognized as important, for example in the Köppen system (Bailey 1996), was largely unexplored.

Recent research has begun to shed light on the relationships between precipitation and vegetation. I will first summarize the broad patterns of climate and vegetation distribution over North America with emphasis on the conterminous United States. Several regions will then be highlighted as case studies for the development and examination of hypotheses concerning the importance of precipitation seasonality on vegetation distribution. These hypotheses will be examined using new models of vegetation distribution and dynamics, with particular focus on a few of the case study regions. Future scenarios of climate and vegetation change will then be examined through known mechanisms and past excursions of climate and vegetation. Finally, a few thoughts on current uncertainties and future directions will be discussed.

Overview of North American Climate and Vegetation Patterns

Although the Köppen and Holdridge models and others provided strong correlates between climate and vegetation, it was recognized that knowledge of direct causation remained elusive. Understanding causation is important for gaining confidence in future projections of possible changes in the biosphere due to climatic change. Progress toward understanding the relationships between climate and vegetation was significantly advanced when Reid Bryson and several colleagues and students recognized that vegetation zones appeared to correspond to the seasonal and spatial distributions of air mass regions (Bryson 1966; Bryson and Hare 1974; Wendland and Bryson 1981). Air mass regions are in general circumscribed by the seasonal dynamics of jet stream patterns and their interplay with continental margins and mountain ranges. The arctic, polar front, and subtropical jet streams are of primary importance for delineating air mass regions in North America and the Northern Hemisphere in general. The Rocky Mountains and the Himalayas exert the strongest influence on jet stream patterns of any mountain ranges and thus produce relatively clearly defined air mass regions and associated biotic regional patterns in their immediate proximity (Tang and Reiter 1984).

Bryson and colleagues pointed out that the southern and northern limits of the Boreal Forest correlate with the summer and winter positions, respectively, of the Arctic Front. They also delineated other air mass and vegetation patterns in the eastern United States. Mitchell (1976) analyzed air mass regions in the western United States in relation to the polar front and subtropical jets. Neilson and Wullstein (1983) suggested direct causal relationships between Mitchell's air mass gradients and the distributions of two oak species in the Rocky Mountain and Colorado Plateau regions. Wendland and Bryson (1981) extended the air mass analyses to the mapping of the general surface airstream regions of the world.

The North American air mass regions were recently depicted by Hayden (1998), another colleague of Bryson's, who related their disposition to the distributional patterns of prairie regions. Hayden's regions II and IV in the Northwest are bound to the north and on the south by the average summer and winter positions of the polar front jet stream (fig. 4.1). Cyclonic storms are steered into the Northwest in the winter and into Canada in the summer, defining the winter-wet, summer-dry character of the region, which correlates with the distribution of temperate conifer forests in areas with sufficient winter precipitation to support forests. Regions III and V in the Southwest are similarly dry in the summer, as they are bound on the north by the polar front and on the south by the subtropical jet stream. However, because regions III and V are south of the average polar front position, they receive less winter precipitation on average and support a variety of more arid ecosystems, ranging from savannas through chaparral to desert vegetation. Variable excursions of the polar front south of its average winter position bring episodic winter rains into the Southwest (during El Niño), sometimes in interaction with the subtropical jet stream. These winter rains can support woody vegetation by recharging deep soil water. In warmer regions, such as California's central valley, winter rains can support grasslands.

The subtropical jet circumscribes the Bermuda High, one of three Northern Hemisphere, quasi-stable, subtropical high-pressure cells, and brings summer rains, known as the Arizona monsoon, into region VI from both the Pacific Ocean and the Gulf of Mexico. These summer rains are critical for the establishment of woody vegetation and the maintenance of grasslands (Cable 1975). Another summer moisture regime is defined in regions II and III along the west coast and delineates the influx of summer humidity into the continental margin as far as the Cascade and Sierra Mountains, even though summer rains are sparse in those regions.

East of the Rocky Mountains, the winter and summer positions of all three jet streams, the Arctic Front, Polar Front, and Subtropical, interact to form a complex mosaic of unique air mass regions. The subtropical jet stream in the summer extends diagonally across the Southwest and crosses the Rocky Mountains, defining a zone of summer precipitation along its margin and to

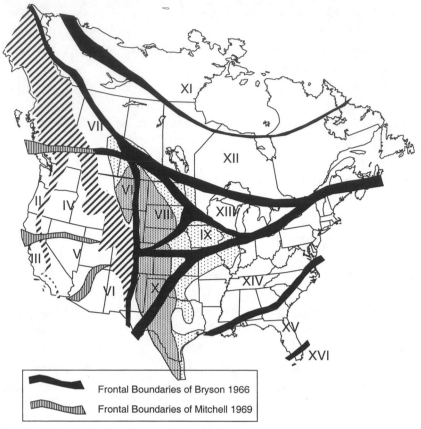

Figure 4.1. The distribution of seasonal air mass boundaries. Stippled areas, east of the Rocky Mountains, indicate the general locations of short-grass and tall-grass prairie regions. See text for explanation of numbered regions. Adapted from Hayden (1998). Used with permission.

the south. Region IX, for example, is closely associated with the "cornbelt" or "Prairie Peninsula" region of the Great Plains and is characterized by a summer-wet, winter-dry precipitation pattern. The subtropical jet shifts south and east in the winter to separate regions XIV and XV, defining the subtropical forests in the Southeast and the more transitional tropical vegetation in Florida. Interannual variability in jet stream position, often determined by sea-surface temperature anomalies, such as the El Niño/La Niña cycles, can shift the jets in any given season to positions more typical of a different season (Neilson 1986). Although the relations between regional air mass and vegetation properties are instructional, the question of causation is still elusive with respect to vegetation distribution and dynamics.

Case Studies and Hypotheses

I will focus on just a few of these regions to demonstrate some of the complex seasonal interactions between vegetation and climate that help to determine the distribution of various vegetation types. Of particular interest are regions IX, V, and VI. Region IX encompasses the "Prairie Peninsula" or "cornbelt," a summer-wet, winter-dry region, and has long been of interest to biogeographers. The region receives about the same annual precipitation as the forested region to the east, but supports only natural savanna or grassland. Yet, most models of vegetation distribution populate the "peninsula" with forests due to the high annual average precipitation. Regions V and VI span the distribution of the "Arizona monsoon," the July-August rains that blanket much of Arizona and New Mexico, but tail off to the north in Utah and Nevada (region V). These western regions bound unique floristic zones and have produced some fascinating and often contentious issues in vegetation distribution.

R. H. Whittaker (Whittaker and Niering 1965) and W. Merriam (1890) both published seminal works on the vertical distribution of vegetation on mountains in Arizona and drew analogs between elevational and latitudinal vegetation zonation patterns. However, such analogies fail to fully recognize the importance of the Southwest summer rains in Arizona in determining the elevational vegetation zones and their regional context. Both the Prairie Peninsula and the Southwest regions are intriguingly related to the seasonal dynamics of the Bermuda High, centered off the southeast coast of North America. We will explore some of these interactions between precipitation seasonality and vegetation distribution using simulation modeling.

Simulating Vegetation Distribution

The role of precipitation seasonality in controlling vegetation distribution will be explored from the perspective of current theories and using a process-based vegetation biogeography model. The Mapped Atmosphere-Plant-Soil System (MAPSS) was explicitly designed to test mechanistic hypotheses of broad-scale vegetation distribution patterns in response to amount and seasonality of precipitation and other climate variables (Neilson 1995). Although many of the air mass dynamics previously described may ultimately control vegetation distribution, the model was constructed from a local perspective with a focus

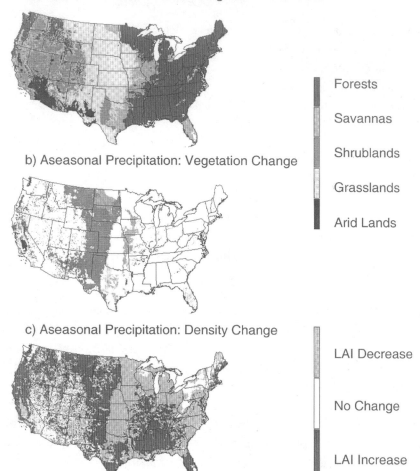

a) Normal Climate: Simulated Vegetation Distribution

b) Aseasonal Precipitation: Vegetation Change

c) Aseasonal Precipitation: Density Change

Forests

Savannas

Shrublands

Grasslands

Arid Lands

LAI Decrease

No Change

LAI Increase

Figure 4.2. (a) The distribution of vegetation in the conterminous United States as simulated by the MAPSS biogeography model (Neilson 1995). The 45 vegetation types simulated by MAPSS have been aggregated for easier display. (b) Changes in the vegetation distribution simulated by MAPSS if annual precipitation were evenly distributed across all months. Shown are the different vegetation types that could occur if the precipitation were aseasonal. (c) Simulated differences in vegetation density (LAI) between the aseasonal and normally seasonal precipitation simulations.

on the proximate causes of vegetation growth and survival. If the concepts are correct, then accurate vegetation distributions and their relations to air masses should be an emergent property of the model (fig. 4.2a).

There are two fundamental modules in MAPSS: a process-based leaf area index calculator (LAI, the area of leaves per unit ground area) and a phys-iologically conceived rule-base for determining leaf form, such as broadleaf versus needleleaf (microphyllous), or evergreen versus deciduous. The LAI calculator determines the seasonal density of the vegetation, which allows the broadscale designation of forests, savannas, shrublands, grasslands, and des-erts. LAI is either constrained by lack of energy (low temperatures) or lack of water and is maximized at a site to just utilize available soil moisture, based on the equilibrium theory of LAI (Woodward 1987). The rule-base determines the kinds of vegetation life-forms that could live at a site, based largely on various seasonal climate patterns and thresholds. For example, if the energy available during the frost-free season is insufficient to support a deciduous plant, then an evergreen is selected.

The vegetation in MAPSS is partitioned into an overstory woody layer that competes with an understory ephemeral layer for light and water. The under-story "grassy" layer is rooted in a surface soil layer, whereas the woody over-story can withdraw water from either the surface or a deeper layer, consistent with the two-layer hypothesis for the existence of savannas (Walter 1971). MAPSS is driven by mean monthly climatology (temperature, precipitation, humidity, and winds) and is calibrated against observed LAI and runoff data. Because the LAI is limited by available soil water, it was deemed necessary to include an accurate, process-based site hydrology, calibrated to stream gages capable of capturing the influence of seasonal precipitation patterns on verti-cal soil moisture content (Neilson 1995).

The MAPSS model was modified and hybridized with the CENTURY bio-geochemistry model (Parton et al. 1988) to simulate dynamic vegetation change. The new model, MCI, is among the first generation of dynamic general vegetation models (DGVMs) (Daly et al. 2000). MCI, like MAPSS, also limits vegetation density based on energy and water with woody plant-grass competition over a multi-layer soil. Unlike MAPSS, the fire model in MCI incorporates state-of-the-art fire simulation processes and determines when and where a fire will occur as well as the level of impact on the ecosystem (Lenihan et al. 1998). MCI accurately simulated the extreme 1910 fire year as well as the Yellowstone fires during 1988 (Bachelet et al. 2001). At present, MCI simulations are based on dynamic natural vegetation; land-use practices, such as conversions of forests and grasslands to agriculture and the wide-spread suppression of wildfires during the twentieth century, are not included in the model.

The Prairie Peninsula

The Prairie Peninsula region (region IX) receives about the same amount of total annual precipitation as the neighboring forests to the east in regions XIII and XIV. However, the precipitation is strongly seasonal over the peninsula, whereas it is rather evenly spread throughout the year over the eastern forested region. The very high summer precipitation is drawn primarily from Gulf of Mexico moisture brought north by the counterclockwise circulation of the Bermuda High. Neilson (1991) analyzed seasonal rainfall and runoff characteristics along an east-west transect through the peninsula. Although rainfall over the forested region is high all summer, runoff steadily decreases through the growing season, yet never disappears by summer's end. However, summer runoff virtually disappears in the Prairie Peninsula region, even though summer rainfall is higher on average than in the eastern forests. The lack of significant, regular winter precipitation apparently prevents the moisture recharge of deep soil layers, which feeds base flow runoff and woody plant transpiration during the warm seasons. Deep soil water appears to be necessary to support closed forests in this region and throughout North America, even when summer rains are high. Neilson et al. (1992) defined a rule-based model that partitions summer and winter precipitation regimes and was among the first to accurately simulate the general outlines of the Prairie Peninsula, lending some support to the two-layer hypothesis and plant competition theory.

The hypothesis for the Prairie Peninsula region (Borchert 1950; Coupland 1958; Manogaran 1983) is that summer rains, on average, could support a closed forest. However, the summer rainfall over the Prairie Peninsula is extremely variable from year to year and decade to decade, because of variation in the strength and position of the Bermuda High (Norwine 1978). Thus, given the low winter precipitation, a failure of the summer rains would stress the forests, causing some drought-induced dieback. The reduction in tree density would allow the grasses to increase in density, which could further impact woody plant recruitment and survival of extant seedlings, saplings, and mature trees. Regular summer rains in the region can build considerable grass biomass and foster abundant fires, which would further prevent the westward extension of the prairie/forest ecotone.

Unlike its rule-based predecessor (Neilson et al. 1992), the MAPSS model was initially unable to simulate the Prairie Peninsula (Neilson 1995) (fig. 4.3a). The hypothesis for the density of woody plants in the Prairie Peninsula, as

a) MAPSS Simulation (no 'Prairie Rule')

b) MAPSS Simulation (with 'Prairie Rule')

c) MC1 Simulation (with variable climate and simulated fire)

Forests

Savannas

Grasslands

Figure 4.3. Three different simulations of vegetation distribution in the Prairie Peninsula region. (a) The original MAPSS simulation, lacking a "Prairie rule." (b) MAPSS simulation, including a "Prairie rule," which recognizes the seasonality of precipitation. (c) Simulation by the dynamic general vegetation model, MC1 (Daly et al. 2000), which includes 100 years of historical interannual variability of seasonal precipitation and a process-based fire model.

previously outlined, requires interannual and interdecadal variability in precipitation, which is not a component of the MAPSS climatology. The rule-based model "worked" because it presumed that the forests were primarily dependent on winter rains, and it ignored summer rains as being either insufficient or temporally unreliable. Thus, we conceived a new rule for the MAPSS model as follows.

MAPSS uses a simple monthly temperature threshold to determine the beginning and ending of the frost-free period, a useful, spatially variable definition of the nongrowing season versus the growing season. Since actual transpiration is calculated each month, the proportion that arises from summer rains can be determined. We defined an empirical function that linearly decrements the maximum LAI that can be supported at a site as a function of the dependency of growing-season transpiration on summer rains. The greater the dependency, the greater will be the decrement from the equilibrium LAI in order to reflect the periodic droughts that sweep the area. With the woody LAI thus constrained, the site LAI is again calculated, which results in a greater amount of "grassy" understory vegetation. Sufficient understory fuels trigger a "fire rule," which further reduces the woody LAI, resulting in savanna vegetation. Thus, the elements of the Prairie Peninsula hypothesis are either explicitly or implicitly rendered in the model and are allowed to operate over the entire modeled area. The new "prairie rule" accurately simulated the Prairie Peninsula, lending additional support to the mechanistic hypothesis, which is ultimately based on the seasonality of the precipitation (fig. 4.3b). Interestingly, when applied globally, the rule also captured the wheat-belt of China and the Pampas of Argentina (unpublished data). MAPSS and other biogeography models normally simulate both areas as broadleaf deciduous forest.

MCI, a dynamic model with a process-based fire submodel, is ideally suited to more directly test the Prairie Peninsula hypothesis. Recent simulations over the conterminous United States, using a gridded 100-year monthly time series of observed climatology, were able to accurately simulate the Prairie Peninsula (fig. 4.3c). The fire submodel initiated several fires in the prairie region during the twentieth century, most recently during the 1988 drought, but also during all prominent droughts of the century. Thus, the simulations support the general hypothesis of the Prairie Peninsula and its relation to precipitation seasonality and variability. That is, the average summer-wet, winter-dry seasonality is insufficient for supplying water to deeply rooted trees and sensitizes the region to summer droughts. Thus, tree density in the Prairie Peninsula is reduced by episodic summer drought, opening the canopy for grasses to increase (surface fuel build-up). Fire then completes the process of determining savanna or grassland vegetation in the region.

The Bermuda High: Southeast versus Southwest Vegetation Patterns in the United States

The Southeast forests of the United States are at the same latitude as the Southwest deserts; both are in the general latitude of the subtropical high-pressure cells that, with dry, descending air, create a latitudinal zone of deserts, including the deserts of the southwestern United States and the Sahara Desert. An important question then is why, given the desert latitude, does the Southeast support forest vegetation, while the Southwest is a desert. The answer appears to lie in the seasonal dynamics of the Bermuda High, which oscillates to a winter position off the east coast of North America and to a summer position further west over the Gulf of Mexico. The clockwise circulation of the Bermuda High, in its eastern winter position, routes moisture from the Gulf into the Southeast, providing sufficient winter precipitation to recharge deep soil layers, and thus allows the existence of forest vegetation. If the continent were shaped differently, lacking a water source in the Gulf, the Southeast might be a natural desert, like others at similar latitudes.

The rains often concentrate around the perimeter of the high pressure, where rising air creates convection and can trigger severe storms. As the High moves toward the west in the spring, it sweeps rain into Texas and neighboring Great Plains states. The westward edge of the moving High pauses for a few weeks in an average year, as it abuts the Rocky Mountains (Neilson 1987a). Abruptly, in early July, the edge of the High pops over the mountains and lodges up against the Sierra Madre Oriental, from where the clockwise circulation routes moisture toward central Arizona, notably Tucson and the Mogollon Rim. Moisture from the Pacific, from both the Gulf of California and the intertropical convergence zone, is routed into the Southwest along the subtropical jet stream, where it mixes with that from the Gulf of Mexico. The air column becomes saturated and very unstable, triggering the well-known Arizona monsoon, which brings precipitation into the Southwest from July through September. As the moisture is deflected from west Texas into the Southwest, there occurs a rainfall minimum in Texas and the southern plains during July and August. The seasonal dynamics of the Bermuda High are thus largely responsible for the occurrence of forests in the Southeast and deserts in the Southwest, with grasslands in between.

Southwest Vegetation Patterns: Temporal, Horizontal, and Vertical Change

The Arizona monsoon provides anywhere from 40 to 60 percent of the annual precipitation in the Southwest, largely in the months of July and August. In contrast, spring and fall are relatively dry. Winter precipitation increases with increasing latitude from Arizona to the north, thus making the Southwest a biseasonal precipitation region. There are three kinds of vegetation patterns of particular interest in the region, each with specific dependency on temporal change, horizontal and vertical patterns, or the seasonality of precipitation: (1) Temporal change has produced a recent broadscale shift from grassland to shrubland in the lowlands. (2) Horizontal vegetation zonation is produced by broadscale, regional climatic patterns. (3) The vertical vegetation zonation patterns described first by Merriam (1890) in the San Francisco Peaks near Flagstaff and later by Whittaker and Niering (1965) in the Santa Catalina Mountains near Tucson are also climate-derived, but they carry unique features due to rainfall seasonality that have not been previously appreciated.

Temporal Change: Southwest Desertification and Precipitation Seasonality
Recent and ongoing conversion from grassland to shrubland in the lowlands of the Southwest (desertification) has been variously attributed to overgrazing and climate, with most emphasis being placed on overgrazing (Buffington and Herbel 1965). However, Neilson (1986) presented a complex hypothesis requiring both climate and overgrazing to cause the shift. The gist of Neilson's hypothesis is that establishment of grasses through the sexual cycle requires abundant summer rains, whereas establishment and persistence of shrubs requires both summer rains for establishment and winter rains for persistence. Both winter and summer rains vary from year to year, and the interannual variance structure appears to be related to the large-scale circulation of the atmosphere during multidecade circulation regimes. Neilson suggested that shifts in the interannual variance of seasonal precipitation, which resulted in more wet winters and dry summers, were unfavorable to grass seedling establishment, whereas grazing would have simultaneously removed asexual reproduction, thus favoring shrubs. Building on the two-layer savanna hypothesis (Walter 1971), Neilson (1986) postulated that winter precipitation favors woody vegetation (shrubs) in the Southwest by recharging deep soil layers, which supply transpirational demands during the summer. However, high summer precipitation, which normally recharges only surface soil layers, should favor grasses, especially if dry winters have reduced competition from shrubs.

Neilson noted that there was a temporal pattern to the interannual variability of summer and winter rains that could be related to Pacific Sea surface temperature anomalies and the effects those anomalies had on the jet stream positions and, hence, the storm-tracks bringing moisture to the Southwest. The temporal rainfall patterns appeared to fall into multidecade periods, or circulation regimes. The period prior to about 1900 was one of meridional circulation, during which there were as few as two to four modes to the polar front jet stream with a strong north-south component as it circled the globe. During meridional flow, the jet tends to resonate, or phase-lock, with the continental margins and mountain ranges, producing persistent rainfall anomalies. Under such conditions some regions experience extended high pressure, and drought ensues, while neighboring regions receive above-normal precipitation, possibly producing floods.

Prior to 1900, the Southwest had several protracted multiyear episodes of winter drought, with above-average summer rain. Such a seasonal pattern should have promoted grassland and inhibited the establishment of shrubs. After 1900, with a shift to a more zonal (east-west) jet stream pattern, winter rains increased but were unpredictable from year to year. However, since about 1940, consistent with a shift toward more meridional flow, winter rains became more temporally and spatially organized, showing two- to three-year persistence, and were clearly related to Pacific Sea surface temperature anomalies (Neilson 1986). In years with wet winters, shrubs would have been favored to establish, while dry winters would have produced shrub mortality. Neilson noted that the few years in the twentieth century when the dominant grasses were observed to seed and establish were all years with dry winters and wet summers.

Thus, in the twentieth century, sexual reproduction (seedling establishment) of grasses was likely reduced due to shifts in the interannual and interdecadal variability of the seasonal rainfall pattern, whereas asexual reproduction (suckers) was reduced due to grazing. Meanwhile, enhanced winter rains allowed establishment of the shrubs, increasing the competition for water with grasses. Additional feedback in soil nutrient processes would have accelerated the desertification process (Schlesinger et al. 1990). That is, the establishment of shrubs tends to sequester soil nutrients in the immediate vicinity of the shrub, reducing the soil nutrient suitability in the shrub interspaces for the establishment of grasses, thus enhancing the perpetuation of "desertified" shrublands.

Horizontal Vegetation Change: Southwest Precipitation Seasonality
The horizontal distribution of plant species is generally controlled by two counter-balancing processes: (1) range expansion via dispersal, establishment, and growth, that is, the completion of the sexual cycle; and (2) range limitation or contraction via mortality of adult organisms. It appears that in the

American Southwest and the Colorado Plateau region, the establishment process is generally controlled by summer rains associated with the Bermuda High and subtropical jet stream, whereas adult mortality is often controlled by winter or spring frost associated with the polar front jet stream (Neilson and Wullstein 1983).

Near its northern limits at the polar front jet stream, the deciduous Gambel's oak (*Quercus gambelii* Nutt.) appears to be limited to asexual (clonal) reproduction. The influence of the Bermuda High and its associated Arizona monsoon diminishes from Arizona into Utah along the west slope of the Rocky Mountains. Neilson and Wullstein (1983, 1986) demonstrated the importance of the summer rains to the establishment (sexual reproduction) and maintenance of the distribution of both Gambel's oak and *Q. turbinella* Greene (canyon live oak). Curiously, the northern limits of the evergreen *Q. turbinella* are disjunct and are more closely related to two summer moisture gradients, the Arizona monsoon and the summer influx of moisture from the Pacific in California, than to the polar front (fig. 4.4a, 4.5a).

Nevertheless, adult mortality in both species appears to be related to winter or spring frost, not water-stress. Being deciduous, Gambel's oak is resistant to winter cold, but it is susceptible to late spring frosts, which can cause defoliation and drain root reserves. A lack of summer rains would preclude the recovery of those reserves. In contrast, *Q. turbinella,* being evergreen, is susceptible to frost at any time of the year. Midwinter excursions of cold, also associated with the polar front, can defoliate the evergreen adult clones (Ehleringer and Phillips 1996). A hypothesis, quantified in the MAPSS biogeography model, is that hard winter frost is frequent in the temperate zone and relatively infrequent in the subtropical zone (Neilson 1995). The MAPSS delineation of the temperate-subtropical transition courses along the topographic scarp separating the subtropical Southwest deserts from the higher-elevation temperate Great Basin and just clips the northern limits of *Q. turbinella* near the Arizona-Utah border. Thus, the distributions of both oak species appear to be limited by both the Arizona monsoon (for seedling establishment) and the winter to spring polar front frost limit (adult mortality).

One of the 45 vegetation classes simulated by the MAPSS biogeography model is Shrubland Subtropical Xeromorphic (ssx), which is a near-perfect mapping of the distribution of *Q. turbinella,* even capturing the disjunct distribution between Arizona and California and its relation to different summer moisture gradients (fig. 4.4b). The classification is based on two

a) Distribution of *Quercus turbinella*

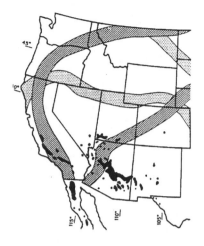

b) MAPSS simulated distribution of
Subtropical Shrubland Xeromorphic

c) MAPSS simulated distribution of SSX
under the CGCM1 future climate scenario

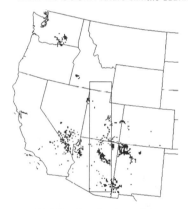

Figure 4.4. (a) The distribution of *Quercus turbinella* in relation to the air mass boundaries from figure 4.1 and described in the text. (b) The MAPSS simulated distribution of Shrubland Subtropical Xeromorphic (ssx), a physiognomic vegetation type that would contain *Q. turbinella* and similar species. (c) The simulated distribution of the ssx vegetation type under the Canadian (CGCM1) future climate scenario, equilibrated to the average simulated climate of 2070–2100. The rectangle coursing through Utah and Arizona shows the location of the latitudinal transect used to analyze seasonal precipitation patterns (fig. 4.5) and elevational ecotones (fig. 4.6).

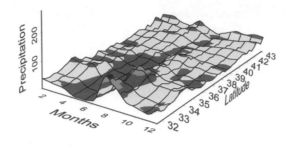

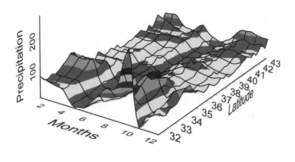

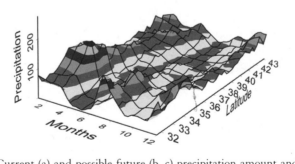

Figure 4.5. Current (a) and possible future (b, c) precipitation amount and seasonality on a latitudinal transect along the west slope of the Rocky Mountains (see fig. 4.4 for transect location). The contour intervals are 25 mm. Two possible future climate scenarios are shown: (b) HADCM2SUL from the Hadley Climate Centre (Johns et al. 1997; Mitchell et al. 1995) and (c) CGCM1, from the Canadian Climate Center (Boer, Flato, and Ramsden 1999; Boer, Flato, Reader, and Ramsden 1999; Flato et al. 1999).

criteria: a water-balance constrained simulation of overstory and understory LAI and a restriction to the subtropical thermal zone due to frost limitation.

The ssx type, however, extends in a ring around the central valley in California, while *Q. turbinella* is limited to the southern part of the valley. Yet, since ssx is a physiognomic classification, one would expect similar species to be located throughout the ssx distribution. Indeed, there are several woody species, in addition to *Q. turbinella,* that appear to occupy this type, notably its close relative, interior live oak (*Q. wislizenii* A. DC.) (Little 1976). Thus, a physiologically based model is able to simulate the distribution of *Q. turbinella* with accurate representation of the constraints of both temperature and water. However, MAPSS, being an equilibrium potential vegetation model, is only able to simulate adult survivorship with respect to frost and water-balance and does not yet simulate the establishment niche.

Vertical Vegetation Change: Southwest Precipitation Seasonality

The vertical zonation of vegetation in Arizona was described first by Merriam (1890) and later by Whittaker and Niering (1965) and has long been touted as an analog of the broad latitudinal vegetation zones of the Earth. That is, the vertical and horizontal vegetation patterns were inferred to be under similar causation and therefore analogs of each other. The logic is that elevational bands of vegetation occur at ever-lower elevations with increasing latitude. Thus, high latitudes are forested, giving way with decreasing latitude to savannas, shrub-steppe or grasslands, and finally deserts, similar in sequence to the vertical zonation of vegetation at lower latitudes. The implication is that elevational ecotones should be roughly parallel and inclined with respect to latitude. However, the unique biseasonal rainfall pattern in the Southwest (fig. 4.5a) appears to alter the vertical zonation of vegetation, creating nonparallel elevational ecotones with latitude. That is, the somewhat simplistic analogy between vertical and horizontal vegetation zonation fails to consider the complexity of the regional air mass gradients and precipitation seasonality (fig. 4.1).

Neilson and Wullstein (1983, 1986), in studying the distribution of Gambel's oak, examined the survivorship of seedlings along latitudinal and elevational climate gradients, encompassing the areas studied by Merriam, and Whittaker and Niering. They found that as summer rainfall diminishes to the north, seedling survivorship became increasingly restricted to more mesic microhabitats and higher elevations where orographic rainfall is higher.

Neilson and Wullstein also documented the role of spring frost in defoliating oaks at their upper elevations.

Based on the Gambel's oak studies, Neilson and Wullstein (1983, 1986) and Neilson (1987b) hypothesized that the upper elevational ecotones should increase in elevation from north to south along the west slope of the Rocky Mountains, being largely under temperature constraints, whereas the lower elevational ecotones, rather than paralleling the upper ecotones, should decline in elevation, or remain approximately level from north to south along the gradient. That is, the presence of summer rains, increasing toward the south, has the effect of shifting the lower elevational limits of vegetation lower than they would be without the summer rains. This is due both to the ability to establish seedlings, given summer rains, and to the improved water balance in general. There is a shift in the controls on ecotones from temperature at the upper elevations to water at the lower elevations. Thus, the interaction between regional gradients of temperature and summer rainfall produces an elevational "wedge" between upper and lower ecotones, narrow in the north and broad in the south. A corollary of the hypothesis is that the vegetation zones described by Merriam and by Whittaker and Niering are not parallel bands, but should be gradually compressed into a common elevation as one moves north along the transect. As the biotic zones become compressed to a common elevation in the north, they tend to segregate in uniquely different microhabitats. Thus, a given topographically complex landscape at the "apex" of the wedge would be spatially more complex in vegetation structure and species composition than a similar landscape in the southern reaches of the wedge (along the north-south transect of the west slope of the Rocky Mountains).

The "wedge" hypothesis was partially tested using the MAPSS biogeography model. Four simulated elevational ecotones were examined along the west slope of the Rocky Mountains: upper tree line, closed forest to open savanna, tree savanna to shrub-steppe, and shrub-steppe to grassland (fig. 4.6a). As hypothesized, the simulated upper elevational ecotone increases in elevation from north to south, while the lower elevational ecotones tend to decrease in elevation from north to south. Interestingly, intermediate and lower elevational ecotones are simulated to increase in elevation with declining latitude in the northern part of the transect and decrease in elevation in the southern part, suggesting complex limits from temperature in the north and summer moisture in the south with variation in total annual precipitation modulating the ecotonal positions throughout the transect. The influence of

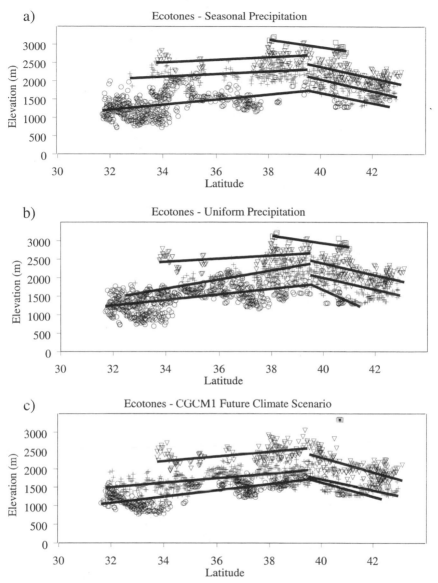

Figure 4.6. Simulated (MAPSS) distribution of elevational ecotones on a transect along the west slope of the Rocky Mountains (see fig. 4.4c for transect location). Ecotones are simulated under (a) observed precipitation seasonality (see fig. 4.5a), (b) uniform seasonality where precipitation is equal across all months, and (c) under the future Canadian climate scenario averaged for the years 2070–2100 (CGCM1, see fig. 4.5c). The upper ecotone is tree line (□). The next is the ecotone between closed forest and open savanna (△), followed by the ecotone between tree savanna and shrub-steppe (+), followed by the ecotone between shrub-steppe and grassland (o).

the Arizona monsoon on the simulated ecotones emerges south of about 39° N latitude (fig. 4.6a) and corresponds to the latitude south of which Gambel's oak seedlings were able to survive the summer (Neilson and Wullstein 1983). Note also the broad overlap of symbols (circles, triangles, and pluses) for the three lower ecotones in figure 4.6a in the northern part of the transect and their vertical separation in the southern part. This mixing of vegetation types at common elevations in the north, yet separation in the south demonstrates the latitudinal shifts in landscape complexity of vegetation distribution as previously hypothesized for this latitudinal transect and as observed in the field (Neilson and Wullstein 1983). These simulated specifics of elevational ecotones along the west slope of the Rocky Mountains comprise a detailed hypothesis, testable with field data and remote imagery.

Aseasonal Precipitation: A Modeling Experiment

Another test of precipitation seasonality hypotheses is to compare vegetation distribution simulated by MAPSS under seasonally uniform precipitation to that simulated under normally seasonal precipitation. Annual precipitation for this experiment was evenly distributed to all months of the year for all locations in the conterminous United States (10 km resolution). All other climate inputs—temperature, humidity, and winds—were left as in the normal climate inputs. Results were analyzed for changes in simulated vegetation distribution and density (LAI).

In general, under uniform precipitation seasonality, locations with normally wet winters and dry summers, such as the West, show increases in vegetation density (fig. 4.2b, c). In contrast, areas with dry winters and wet summers decreased in vegetation density. However, these generalizations are complicated by competition for water between woody and grass vegetation, in the context of total annual precipitation.

As expected, the density of the trees in the Prairie Peninsula increased by up to 50 percent due to a shift of the summer rains to winter, causing a recharge of deep soil layers, thus favoring the trees. Similarly, the prairie grasslands in general were much reduced in grass dominance by an increase of small-leaved shrubs (Great Basin sage) in the short grass region and by broad-leaved shrubs and juniper woodlands in the central plains (aggregated under shrublands and savannas, respectively).

Northwest conifer forests increased in area and density, overtaking savannas in the Willamette Valley in Oregon and parts of the dry interior, due to a shift of overabundant winter rains into the normally dry summer growing season (fig. 4.2b). However, some regions in the West with low but primarily winter precipitation, such as the interior Great Basin and California's Central Valley, shift the competitive balance from shrubs to grasses with a shift of winter precipitation to summer. Incidentally, this shift from shrubs to grasses with the change in precipitation seasonality is consistent with the hypotheses discussed previously regarding the shrubland-grassland temporal shifts in the Southwest.

The wedge hypothesis, previously described, suggests that lower elevational ecotones along the west slope of the southern Rockies are largely under the control of the summer rains, i.e., the Arizona monsoon. If the hypothesis is correct, then shifting to uniform precipitation should move the lower elevational ecotones to higher elevations in the summer rainfall areas. Indeed, under aseasonal precipitation, the tree savanna to shrub savanna ecotone shifted up in the southernmost part of the transect by as much as 700 m with similar but more complex effects on the remaining ecotones (fig. 4.6b).

Future Climates

The responses of vegetation to potential future warming have been examined using the MAPSS model under several future climate scenarios (Bachelet et al. 2001). The future scenarios range over the United States from 2.8 to 6.6°C in temperature increase and about 4 percent to 32 percent in precipitation increase (Neilson and Drapek 1998; Bachelet et al. 2001). The future climate simulations tend to enhance the monsoon rainfall and carry it to more northern latitudes (fig. 4.5b, c). Interestingly, winter precipitation also tends to increase in these scenarios.

Scenarios with increased summer precipitation should favor grasses, while increased winter precipitation would tend to favor woody vegetation. However, as was seen in the uniform precipitation experiment, the outcome of competition between grasses and woody vegetation also depends on the total annual precipitation and the interplay of fire with woody and grass fuels. The competitive outcome can be very complex, but in general tends to favor a large increase in the abundance of grasses in the Southwest desert region, due to an overall increase in precipitation, especially summer rains (Neilson and Drapek 1998; Bachelet et al. 2001).

The simulated evergreen vegetation of the ssx type, i.e., the *Q. turbinella* analog, is sensitive to winter cold and also requires summer rains for establishment (a process currently not simulated by MAPSS nor MC1). With warming, the thermal constraint shifts up in elevation and over the topographic barrier separating the Southwest deserts from the higher Great Basin and thus shifts considerably north through the comparatively flat Great Basin. This topographic release of the thermal constraint, coupled with the northward advance of monsoon rains, allows the ssx type to shift quite far north (fig. 4.4c), as has been documented for its mid-Holocene distribution (Cottam et al. 1959).

As expected, the enhancement of the summer rains under future warming also shifted lower elevational ecotones along the west slope of the southern Rockies to even lower elevations (fig. 4.6c), while shifting the upper elevational ecotones to higher elevations. In effect, the "wedge" of ecotones shifts north. These shifts would produce complex effects on local to regional biodiversity patterns.

The mid-Holocene thermal maximum of about 5000 years ago has often been used as an analog of possible future warming, even though the causal mechanism is different (orbitally induced solar forcing versus greenhouse gas forcing). However, in both the future and mid-Holocene cases, it has been proposed that global warming enhances the subtropical jet stream and the moisture associated with it (Davis 1994). Thus, the Arizona monsoon is thought to have expanded north during the mid-Holocene, assisting in the migration of *Q. turbinella* into northern Utah at that time, a distance of about 400 km (Cottam et al. 1959; Neilson and Wullstein 1983).

The potential, explosive shift of ssx to the north also suggests a hypothesis for the rapid expansion of ponderosa pine (*Pinus ponderosa* Laws.) during the Holocene (Anderson 1989). Ponderosa pine, as with the oaks, requires summer rainfall for seedling establishment (Harrington and Kelsey 1979). It is more cold tolerant than the oaks, so can live further north and at higher elevations. During the early to mid-Holocene warming, the Arizona monsoon would have brought summer rains deeper into the Great Basin, as well as releasing thermal constraints over large areas, due to the flat topography of the Basin. Ponderosa pine could have expanded rapidly from local refugia, shifting from nonexistent in the fossil record to regionally abundant (Anderson 1989). One implication of global warming, then, is that locally rare species could rapidly expand throughout much of the interior West, due to both thermal releases and increases in summer precipitation.

Summary, Conclusions, and Future Directions

The importance of precipitation seasonality to the distribution and function of natural vegetation is realized primarily via the effect it has on the timing and vertical distribution of water in the soil (Goldstein and Sarmiento 1991). Simply put, the two-layer hypothesis of woody vegetation and grass interaction suggests that grasses withdraw water primarily from a near-surface soil layer, while woody vegetation can withdraw water from both the surface and deeper layers. The seedling establishment of many woody plants is also constrained by summer surface soil moisture and competition for that moisture from grasses and other ephemerals. The seasonality of precipitation in conjunction with thermal regimes that govern the seasonality of plant activity determines to what depth water may infiltrate. If precipitation falls during the active growing season, then plants tend to take up the water as soon as it enters the soil in the surface layer, hindering deep percolation and recharge of soil moisture in those deeper layers. However, if the precipitation falls during plant dormancy or cool periods with low evaporative demand, then the water can percolate to deeper layers. Water stored in deeper layers then becomes available to deeply rooted vegetation during the growing season.

The Prairie Peninsula is a region with wet summers and relatively dry winters. Thus, the climate tends to favor grasses, even though the annual average amount of precipitation could support a closed-canopy forest. The summer rains, being largely convective, are inherently variable on both interannual and interdecadal time scales. Failure of summer rains compounds the problems of dry winters, further favoring grasses over trees. In contrast, the Northwest is a summer-dry, winter-wet region. This summer dry area with predictable winter rains favors woody vegetation with deep roots, to the detriment of grasses even in the drier interior West.

The Southwest is biseasonal with strong precipitation peaks in both winter and summer, but with spring and fall dry periods. As in the Prairie Peninsula, however, the summer rains are quite variable from year to year. The winter rains in the region, being comparatively low, are also variable, and are tied to El Niño/La Niña weather oscillations. Thus, the region is in a delicate competitive balance between woody and grass vegetation. Changes in grazing and subtle shifts in precipitation seasonality and year-to-year variability can shift the balance between grasses and shrubs in the region (Neilson 1986).

The two-layer hypothesis also has implications for seedling establishment of woody vegetation. In summer-dry regions of the West, most woody vegetation requires some level of summer precipitation for seedling establishment (Neilson and Wullstein 1983; Weltzin and McPherson 2000). Even if summers are relatively dry on average, short periods of summer precipitation may suffice to get a cohort established within the life span of long-lived woody plants.

The seasonality of precipitation is also critical in modulating the responses of vegetation to future climates. Simulations under future climates can produce nonintuitive results. For example, in summer-dry regions, such as the interior Northwest, lower elevational ecotones of woody vegetation might shift up in elevation under hotter, more evaporative climates. However, in summer-wet areas, notably the Southwest, lower elevational ecotones could shift down in elevation, due to increases in summer precipitation.

Vegetation distribution patterns have horizontal, vertical, and temporal components of change. Understanding the interactions among these three components has grown considerably in recent years with the seasonality of precipitation and its interannual and interdecadal variability being increasingly appreciated. Horizontal and vertical gradients in vegetation patterns are clearly related to regional air mass gradients, producing large-scale patterns of vegetation distribution as emergent properties of local vegetation processes in interaction with local climates. Accurate simulation of these large-scale patterns, therefore, must occur through process-based approaches that capture the detailed local complexity and temporal patterns of vegetation change.

The general success of both the MAPSS and MC1 models in simulating accurate vegetation distributions in the strongly seasonal climates discussed above provides support for the two-layer hypothesis and the importance of competitive interactions between woody and grass vegetation. Nevertheless, uncertainties remain. The models currently do not include an explicit regeneration niche to correctly simulate the shallow roots of woody seedlings and their susceptibility to a lack of summer rains. Thus, some simulated range distributions under both current and future climates could exceed those that would occur with proper establishment limitations. Regeneration algorithms are currently under development. Horizontal heterogeneity of both soil moisture and vegetation distribution becomes increasingly important in drier environments. The models do not currently handle these horizontal processes, but algorithms are being developed. The models are very early in development and will improve as more sophisticated algorithms are incorporated and additional sensitivity analyses can help focus research on areas of uncertainty and high sensitivity. In addition, experimental studies involving seed and seedling transplants and manipulations of precipitation seasonality can help clarify the theories, test hypotheses, and improve the models.

Acknowledgments
I would like to acknowledge Ray Drapek, Oregon State University, for assistance in operating the MAPSS model, Geographic Information System analysis, and in the preparation of figures under funding from the USDA Forest Service (JVA 98-5128). Jim Lenihan and Dominique Bachelet (Oregon State University) provided helpful discussions. Jake Weltzin and Guy McPherson provided very helpful review comments, which considerably improved the paper. Bruce Hayden (University of Virginia) kindly provided figure 4.1.

5

Approaches and Techniques of Rainfall Manipulation

M. KEITH OWENS

Studying the effects of precipitation on arid and semiarid terrestrial eco-systems is difficult because rainfall is temporally and spatially sporadic. An-nual precipitation patterns and spatial variation due to convective summer storms in semiarid communities create logistic problems for studying rainfall. Therefore, precipitation must be regulated at appropriate temporal and spa-tial scales to study how changing precipitation patterns and intensities may affect terrestrial ecosystems. Experiments must either be designed at multiple locations and continued for many years to account for natural variability in rainfall or be conducted in controlled greenhouse environments. When using natural rainfall, discrete precipitation treatments are impossible to investigate and analyses must consider continuous treatment levels. Under greenhouse conditions, precipitation can be controlled, but a host of other environmental and edaphic factors are also altered. An alternative is to control the amount and timing of rainfall through the use of rainout shelters. These shelters prevent natural rainfall from reaching plots, and irrigation water is sub-stituted for rainfall. Multiple temporal and spatial scales must be considered in the design and construction of rainout shelters and supporting irrigation systems. The purpose of this chapter is to discuss the advantages and disad-vantages of various rainout shelter designs in arid and semiarid ecosystems.

Experimental rainout shelters have been used in agricultural research for nearly 40 years, with early shelters described by Bruce and Shuman (1962), Horton (1962), and Hiler (1969). These shelters were large, movable struc-tures; numerous modifications have been made in the design and construction of rainout shelters. The ultimate purpose of rainout shelters, regardless of the modifications, is to eliminate rainfall on an experimental plot with only mini-mal alteration to the microenvironment. Two classes of shelters—movable and fixed—have been developed to accomplish these goals. The remainder of this chapter will discuss the relative merits of these two types of shelters and the construction materials typically used for each.

Fixed versus Movable Shelters

The decision to use either a fixed or movable rainout shelter depends largely on research objectives and cost. Lightweight fixed shelters are typically used when plots need to be sheltered for short periods (e.g., one season to one year) or when research necessitates heavy soil disturbance (e.g., root harvesting). Shelters of this type are typically easy to construct and dismantle, so the annual setup costs are minimized. In contrast, some fixed shelters are built for multiple-year use in long-term studies of native communities (e.g., Svejcar et al. 1999; Fay et al. 2000; Weltzin and McPherson 2000). The effects of such shelters on microclimate must be considered. The cumbersome structural materials associated with large fixed shelters may mandate inclusion of an ambient precipitation treatment beneath a rainout shelter and an unsheltered control (Fay et al. 2000). Plant response to the microenvironment caused by the shelter could then be isolated from the precipitation effects. An alternative approach is to monitor environmental conditions under the shelter and in an open area (Svejcar et al. 1999), which facilitates the interpretation of results but does not allow statistical separation of the shelter effects and experimental treatments. Movable shelters affect the microenvironment only while they are located over the plots and may be particularly appropriate for long-term studies. The high cost of building movable shelters virtually necessitates long-term, or multiple, experiments to amortize the expense.

Fixed Shelters
Fixed shelters vary in size from those that shield individual plants to those that shield small natural communities. Jacoby et al. (1988) used subcanopy rainout shelters to limit rainfall reaching the rooting zone of mature honey mesquite *(Prosopis glandulosa)* trees. The shelters were low (0.6 m tall) and did not interfere with the light environment under the tree canopy, but did prevent rainfall from reaching the rooting zone. These simple shelters were composed of a wooden superstructure, plastic sheeting to intercept rainfall, and poultry netting to support the plastic. Reynolds et al. (1999) constructed

shelters over individual shrubs to exclude rainfall during certain times of the year. The shelters were sufficiently tall to allow normal wind movement and therefore maintain air and surface soil temperatures beneath the shelters near ambient temperatures. Shelters were uncovered during winter to allow normal precipitation.

Svejcar et al. (1999, and chapter 6, this volume) developed fixed rainout shelters with permanent roofs to study the effects of precipitation seasonality on native sagebrush (*Artemisia tridentata*) communities in Oregon. These rainout shelters were composed of pole-and-truss construction and a fiberglass roof. The wooden superstructure and fiberglass reduced photosynthetically active radiation by 50 percent, so the authors recommend including an unsheltered control plot to detect shelter-induced changes in microclimate.

Movable Shelters

Movable shelters have generally been large structures that can accommodate multiple replications of precipitation treatments. Shelters have ranged in size from 266 m² (slightly smaller than the fixed shelter of Svejcar et al. [1999]) to 1300 m² (Martin et al. 1988). Originally, movable shelters were designed to accept conventional farm machinery in agronomic research. Smaller shelters for use in remote field sites have been developed recently and are currently being used at the Buxton Climate Change Impacts Laboratory in the United Kingdom (Grime, pers. comm., 3 December 1999) and in semiarid savannas of Texas (Owens, unpub. data; fig. 5.1). The supporting structure and rail system is constructed from angle iron, shelter roofs are built with angle aluminum frames and moved with 12-volt direct-current motors, and total plot size is 3 m by 3 m. Shelters of this size are appropriate for single precipitation treatments, which require multiple shelters and allow an interspersion of treatments in a native plant community. The impact of movable shelters on the microenvironment is largely a factor of the type of roof material, as discussed below (see "Roofing Materials," below).

Cost

Rainout shelters described in this review were constructed between 1980 and the present, and unfortunately the cost of building each structure was not always included in the original manuscript. To compare relative costs of different designs and sizes, it is necessary to estimate construction costs at a baseline date. Therefore, all the costs have been adjusted to 1999 U.S. dollars

Figure 5.1. Movable rainout shelters used in native honey mesquite *(Prosopis glandulosa)* communities in south Texas.

using U.S. Bureau of Labor price indices; costs reflect materials but do not include labor.

Costs for fixed shelters are readily calculated because standard building materials and practices are used. Fixed shelters are obviously less expensive to build than movable shelters because a track and drive system are not needed. Small shelters built beneath the canopies of mature trees, similar to the shelters built by Jacoby et al. (1988), would cost about $1,000 each, while the large shelters (12 m by 30 m) described by Svejcar et al. (1999) would cost about $22,000 each. If changes in the microenvironment do not affect the research objectives, the lower cost of fixed rainout shelters make them appealing.

Movable shelters are significantly more expensive and much more variable in cost than fixed shelters. Materials for large corrugated iron shelters such as the shelter used by Ries and Zachmeier (1985) cost about $82,000 in 1985 but would have cost over $146,000 in 1999. On a smaller scale, the materials for the movable shelter designed by Kvien and Branch (1988) would cost about $3,200, and materials for the shelters used in our research (3 m by 3 m) would cost about $1,250. A critical factor in using smaller shelters is that a single treatment usually is applied under each shelter. If an experiment employs five treatments and three replications, a total of 15 shelters are needed and material costs would approach $20,000.

Design Considerations

Fixed shelters and movable shelters are structurally similar in terms of support structure and roofing materials. The obvious differences are in the tracks, drive mechanism, and control system. Foale et al. (1986) reviewed the construction of movable shelters and identified six major components: site, tracks, structure, drive, power supply, and controller. Their review was comprehensive through 1986, but focused on large shelters used in agronomic situations. This section will focus on new designs and materials available for shelters in ecological and agronomic research. Specifically, this section will review the site and size of shelters, the superstructure support, roofing materials, controller and drive mechanisms, belowground water management, and irrigation systems under the shelters.

Site and Size

The research site can have an overriding effect on the shape and size of the rainout shelter. Shelters used in agronomic research typically are placed on relatively flat land used for production agriculture, so the site may have a minimal effect. In contrast, the architecture of plants or topography of plots may affect shelter design in natural communities. When sampling on a slope, there are compelling arguments both for orienting the shelter across the slope and for orienting the shelter perpendicular to the slope. Movable shelters oriented across the slope do not have a gravitational handicap to overcome, so the shelter is easier to move. Positioning the shelter this way also decreases the possibility of the parked shelter affecting the plot. Orienting the shelter perpendicular to the slope and having the parked shelter on the upslope side allows the shelter to move over the plot using gravity and thus does not require a drive mechanism to cover the plot. However, the shelter must be moved away from the plot by hand after the rain stops, so the plots cannot be far from the research headquarters. A potential problem with parking the shelter in the upslope position is exposure of the area normally covered by the shelter. The newly exposed area will be hydrologically different than the rest of the community, with decreased infiltration and increased splash erosion and runoff. Unfortunately, the increased runoff is aimed downhill at the research plots. Positioning the shelter on the downslope side of the plots can mitigate most of these shelter effects, but the effort to move the shelter over the plot is increased by the downhill gravitational pull. This obstacle can be overcome through careful engineering, as shown by J. P. Grime and J. R. Tippets (pers. comm., June 1997), who developed a counterweight system that worked with small movable shelters to allow parking them on the downhill side of the plots. An alternative on slight slopes is to increase the power of the drive system to move the shelter (Heitschmidt et al. 1999).

Most recent modifications to rainout shelters have adapted the size and shape of the shelter to the particular experimental design and plant being studied. Large shelters are needed to accommodate farm machinery in agronomic research plots and to accommodate mature natural communities. The upper limit on shelter size is imposed primarily by weight and cost. Upchurch et al. (1983) constructed a shelter that covered 890 m². The shelter consisted of two opposing buildings that moved toward one another when activated. Martin et al. (1988) also used a dual-structure design to create a shelter that covered 1300 m². Each of the buildings used by Martin et al. weighed about 7.5 metric tons and was moved by a five-horsepower, three-phase (480 volt) motor. Other shelters are built as single buildings and cover a proportionally smaller area. Johnson and Rumbaugh (1995) used a single shelter covering about 470 m² to study native and introduced perennial grasses. A single structure design also was used by Ries and Zachmeier (1985, 348 m²), Lewin and Evans (1986, 266 m²), and Larsen et al. (1993, 384 m²) in North Dakota, New York, and Montana, respectively. Each of these shelters included modifications for high winds (discussed in the "Roofing Materials" section) and frozen soils (discussed in the "Track Structure" section).

An additional consideration in determining size of a rainout shelter is sampling design and the interspersion of treatments for the experiment. On sites which are vegetationally and edaphically homogeneous, a single large shelter may allow ample space for treatment replication (e.g., Larsen et al. 1993; Heitschmidt et al. 1999; Svejcar et al., chapter 6, this volume). For example, the size of shelters used by Svejcar et al. (1999) was determined by calculating an adequate sample size of sagebrush plants for each treatment: to include 10 experimental shrubs, a plot size of 7.5 m² was needed. Each shelter was 12 by 30 m, which allowed three precipitation treatments and adequate borders beneath the shelters. They constructed five shelters to ensure adequate statistical replication. On highly diverse sites, proper replication may require more small shelters. If experimental treatments are likely to affect soil water content through subsurface flow, then small shelters in which a single treatment is applied under each shelter may be most appropriate to prevent confounding the treatments (Reynolds et al. 1999; Fay et al. 2000). In any case, statistical requirements should be balanced with cost and logistical concerns associated with construction and maintenance of multiple shelters.

Superstructure Support

Fixed and movable shelters have been constructed from a variety of building materials, including wood, aluminum, and steel. Wood has been used extensively for fixed shelters (Jacoby et al. 1988; Bittman et al.1987; Reynolds et al. 1999; Svejcar et al. 1999) but is also used for movable shelters (Larsen et al. 1993). Simple pole-and-truss construction can span distances of 12 m in both fixed (Svejcar et al. 1999) and movable (Larsen et al. 1993) shelters. Wooden supports need to be larger than metal supports and will therefore block more light, which is particularly important for fixed shelters in which changes in the microenvironment may alter experimental conditions.

A primary consideration for movable shelters is the weight of the roof and supporting structure. This weight has direct implications for the strength of the supporting rails and the size of the motor needed to move the shelter. One of the lightest movable shelters was designed by Hatfield et al. (1990), who used steel cables stretched over experimental plots. Heavy-duty vinyl was threaded onto the cables using ceramic insulators, and a 12-volt automotive starter motor was used to pull the vinyl over the plots. This design allowed the researchers to move the rainout shelter between and within seasons to cover different plots. Other lightweight structural parts include electrical conduit and angle aluminum. Electrical conduit used by Clark and Reddell (1990) supplied insufficient strength, was destroyed in a windstorm, and was replaced with structural steel during the first growing season. The angle aluminum used in our rainout shelters has withstood light windstorms but has not experienced any major storms yet.

Tubular steel construction has been used on several fixed shelters. Weltzin and McPherson (2000) used a tubular steel frame to modify the seasonality of precipitation beneath fixed shelters in Arizona. These shelters were relatively small (16 m by 4 m) and did not require any additional structural support. A larger fixed shelter was constructed by Fay et al. (2000) in Kansas using a commercially available greenhouse design with a galvanized tubular steel frame.

Most rainout shelters used in agronomic research include structural I-beams in the design. The dual structure rainout shelter described by Upchurch et al. (1983) used steel I-beams as the major roof girders, which allowed the building to span 18 m. Lewin and Evans (1986) adapted the steel frame of a commercial greenhouse for their movable shelter. The obvious drawback to these designs is their great weight. The dual shelters used by Martin et al. (1988) each weighed 7.5 metric tons, which complicates the drive mechanism, rail support, and wind hold-down structure.

WIND HOLD-DOWN TECHNIQUES.   Regardless of the design of movable shelter, the roof assembly must be fastened to prevent wind damage (Upchurch et al. 1983). Even a slight shift in the position of the roof can cause the shelter to fall off its tracks. Most hold-down devices are simply metal pegs attached to the wheel assembly that hook under the stationary rail. A depth adjuster on the hold-down is necessary to minimize the distance between the rail and the peg to prevent minor movements. An alternative is to use wheels on the side and bottom of the rail. The wheels remain in contact with the rail at all times and guide the moving roof assembly. For this assembly to remain effective, the tracks must be extremely rigid and unmoving, otherwise the roof will bind and short-circuit the motor or jump the track.

TRACK STRUCTURE.   Design of the track structure depends on environmental conditions and weight of the roof. In areas susceptible to freezing soils, the footings for the tracks need to be buried below the frost line to prevent movement. On a shelter constructed in Mandan, North Dakota, footings were 1.5 m deep to avoid frost damage (Ries and Zachmeier 1985). With footings this deep, the tracks remained parallel and level. The foundation wall used in this construction also isolated plots from subsurface water movement. A more typical foundation is similar to a pier, in which support posts are set in concrete at regularly spaced intervals. Depending on the weight of the roof, the supports are usually 3 to 4.5 m apart. On smaller shelters, the track structure may be supported only at the ends.

Environmental conditions, especially wind direction and snow load, influence track design. For large shelters there are certain advantages to orienting the tracks parallel to the prevailing wind direction (Ries and Zachmeier 1985), although there are also disadvantages. An advantage is that wind resistance is decreased on the sides while the shelter is at rest, and consequently there is less likelihood of forcing the shelter off its tracks. A corresponding disadvantage is that a larger motor may be needed to move the shelter into the wind. The dual structure design of Upchurch et al. (1983)

offset this problem, because as one shelter moved into the wind, the other shelter moved with the wind. The force of the wind on a rainout shelter is a function of the size of the side and end walls, and the slope of the roof (Foale et al. 1986). Load factors for several different shapes of shelters are calculated by Foale et al. (1986).

A second advantage of orienting the tracks parallel to the prevailing wind is in snow management. If the shelter is parked on the downwind side of the plots, the wind will keep the shelter clear of snow and the snow will not drift on the plots (Ries and Zachmeier 1985). When snow loads may be a problem, extra attention to the placement of footings may minimize the potential of storm damage. Larsen et al. (1993) designed their shelter such that a footing was under each of the support posts for the roof when the shelter was in the open position. The roof could then support considerable weight with a low likelihood of collapse.

The total length of tracks should be twice the length of the shelter plus a buffer area between the plots and the parked shelter. For large shelters, the total track length can approach 60 to 75 m. Tracks on large shelters are usually structural I-beams in which the wheels of the roof ride on the top of the beam and the hold-down devices fit under the flange of the I-beam. Inverted railroad rails have also been used as tracks. On smaller shelters, the tracks can be simple angle iron, positioned either with a flat side up or inverted so a grooved tire can ride on the track. Several small shelters have been designed without a track system. The cable system described by Hatfield et al. (1990) does not require a track because of the lightweight vinyl covering. Another relatively small shelter (8.6 by 7.2 m) not requiring an in-ground track system was described by Kvien and Branch (1988). This shelter was supported by automotive tires that rested directly on the ground. The tires were guided by pvc irrigation pipe used as bumper rails. The advantage of these small shelters is that they are easily moved from one site to another, which avoids problems associated with soil disturbance and crop rotation. Plots still must be trenched and lined to avoid subsurface soil water movement (See "Belowground Water Management," below).

Elevated rails have been used to decrease the total weight of the roof structure. The main vertical supports can be eliminated from the moving portion of the shelter and transferred to the stationary portion of the structure. This design decreases the energy requirements for moving the shelter, but elevated rails may be susceptible to soil movement, which can cause rail misalignment (Upchurch et al. 1983). When the rails shift, the shelter may fall off the track or become stuck. Elevated rails therefore require a secure footing that is unlikely to shift as the soil settles.

Roofing Materials

The type of roofing material is dependent on the type of shelter and the objectives of the experiment. The main function of the roof is to divert rainfall either to a drain system or to a temporary storage facility, so the water can be reapplied later. The roof also must withstand wind, heat, snow, and long exposure to direct light. Most movable shelters have been clad in a durable opaque sheathing such as corrugated iron, whereas fixed shelters have been designed to transmit as much light as possible.

The lightest and least expensive roofing material is clear plastic sheeting. This is particularly appropriate where the shelter is designed for temporary use during the growing season (Reynolds et al. 1999; Fay et al. 2000), where the roof is moved over a plot by hand (Bittman et al. 1987), or where it is replaced annually (Lewin and Evans 1986). The major disadvantages of plastic sheeting are sagging due to heat and wind, and reduced light transmission from dust accumulation and photodegradation. The problem of excessive sagging can be alleviated by including more purlins (horizontal supports) in the superstructure or by reinforcing the plastic with a fine mesh wire such as poultry netting (Jacoby et al. 1988), although either strategy decreases light transmission. Dust accumulation is often a problem with any of the roofing materials but is particularly problematic with plastic, because it can be difficult to clean without tearing. Alternatives to 6-mil plastic sheeting may be heavy-duty vinyl (Hatfield et al. 1990) or uv-transparent polyethylene greenhouse film (Fay et al. 2000). In fact, Hatfield et al. (1990) used vinyl sheeting over a three-year period without replacement. Plastic sheeting does affect the microclimate under the rainout shelter. While air temperatures remain close to unsheltered air temperatures (within 1°C), total daily radiation may be reduced by 20 to 35 percent, and soil temperatures at 2.5 cm may be significantly higher (up to 6°C) under the shelter in arid ecosystems (Reynolds et al. 1999). In a more mesic ecosystem, plastic sheeting resulted in a 22 percent decrease in radiation, no difference in air temperature, and only a 1.2 to 1.8°C difference in soil temperature between sheltered and unsheltered plots (Fay et al. 2000).

Corrugated fiberglass and clear polycarbonate panels are durable roofing materials. Transparent corrugated fiberglass transmits about 70 percent of the total radiation but only about 50 percent of the photosynthetically active radiation (Svejcar et al. 1999). White or translucent fiberglass also transmits about 50 percent of the photosynthetically active radiation. Kvien and Branch (1988) reported that fiberglass transmitted about 90 percent of the ambient radiation when the fiberglass was new, but only about 70 percent when the fiberglass was three years old. Reduced light transmission associated with aging fiberglass may necessitate periodic replacement of roof panels.

Semirigid polycarbonate sheets have been used on fixed shelters for about 15 years. Clark and Reddell (1990) installed polycarbonate panels in 1985 that were rated for 85 percent light transmission but that reduced light by 30 to 40 percent. Higher light transmission was possible only when the panels were new, clean, and the light was incident to the panels. Rapid discoloration of the panels and low light transmission prompted the authors to replace the panels after one year. They recommended that corrugated fiberglass be used in rainout shelters. On the other hand, Svejcar et al. (1999) started with corrugated fiberglass and switched to polycarbonate panels. New polycarbonate panels are rated at 92 percent light transmission by the manufacturer, with a greater transmission during early morning and evening hours than fiberglass. The shelters transmitted about 75 percent of the photosynthetically active radiation, reduced wind speed an average of 28 percent, reduced relative humidity by 1.5 percent, and increased soil and air temperatures by 16 and 4 percent, respectively (Svejcar et al. 1999). The wooden superstructure used by Svejcar et al. (1999) may have reduced light, especially during the morning and evening when the sun was low in the horizon.

Corrugated iron appears to be the preferred roofing material for large movable shelters (Teare et al. 1973; Upchurch et al. 1983; Ries and Zachmeier 1985; Martin et al. 1988; Larsen et al. 1993). This is obviously the most durable material, but it is also the heaviest, and therefore requires a substantial foundation, superstructure, and power train. Corrugated iron has obvious advantages: it is longer-lasting than materials discussed previously, it resists damage from hail and other violent storms, and it is readily available. A disadvantage, however, is that it does not transmit light, so the shelter cannot remain over the plot for long time periods. Dugas and Upchurch (1984) measured the microclimate under a dual-structure rainout shelter and found that air temperature increased by 6°C within an hour of covering the plots, that wind was essentially eliminated, and total radiation was reduced by 40 percent. Direct radiation was obviously eliminated. To avoid drastic changes in the light environment, especially changing the photoperiod, Martin et al. (1988) constructed a shelter with about 25 percent of the roof covered with fiberglass panels.

Controller and Drive Mechanisms

The simplest movable shelters are moved by hand. Bittman et al. (1987) constructed a plastic-covered shelter that required moving the plastic over the plots whenever rain was imminent. They constructed 12 shelters and each shelter required 1 to 2 minutes to cover. Under the best of circumstances, 15 to 20 minutes were required from the time the first plot was covered until the last plot was covered. Covering the plots took longer if the wind exceeded 4.2 m s^{-1}. Clawson et al. (1986) also used shelters that were manually covered when rain was predicted. If rain fell during the night, or at other inconvenient times, the plots were not covered. A more substantial manual shelter was used at Brookhaven National Laboratory, where a pickup truck was used to pull the shelter over the plots when rain was expected (Lewin and Evans 1986). This shelter was later modified to include a dedicated drive mechanism.

The size and type of motor required to move a rainout shelter depends largely on the weight of the shelter and the proximity to electricity. Small movable shelters can be powered by a 12-volt direct-current motor, such as an automotive starter motor (Hatfield et al. 1990) or an automotive windshield wiper motor (J. R. Tippetts, pers. comm., 2 December 1999). Both of these motors deliver sufficient torque to move a shelter, although the starter motor must have a gear reducer to decrease the speed. These motors are limited to fairly light shelters, so they are ideal when shelters cover a single treatment and are interspersed in a study area. Because the shelters are light, the motor can be attached directly to a drive wheel (fig. 5.2) and a complicated cable system or a gear and cog drive system are not needed. When electricity is not readily available, as is the case for most remote field sites, the 12-volt direct-current motor running off a car battery and solar panel may be the only way to construct a movable shelter.

Large shelters require a permanent source of electricity to run alternating-current motors. Generally one motor can power the shelters, although separate motors are sometimes used in dual-structure designs. Motors generally range from 1 to 3 horsepower, although 5 horsepower motors have been used on the heaviest shelters. The drive mechanism is more complicated than that of small, light shelters because of the energy requirement to start moving the shelter and the momentum of the shelter once it is moving. A slip clutch is necessary between the motor and the shelter to reduce the strain on the motor when the shelter is first activated (Ries and Zachmeier 1985; Martin et al. 1988). This allows a gradual increase in speed until the shelter reaches its maximum speed. When the shelter has reached the end of its travel, it must

Figure 5.2. Drive mechanism for small rainout shelters including the rail, drive wheel, 12V DC motor, and a limit switch.

stop gradually to avoid excessive strain on the entire system. A limit switch can be set to shut off electricity before the shelter reaches the end of the track to allow room for coasting to a stop.

Most shelters control the length of movement by using limit switches. These switches are located at the end of the track on both ends of the rail and are designed to turn off electricity when triggered. In addition, a safety switch should be included to prevent a runaway shelter in case the limit switch does not operate properly. The safety mechanism uses redundancy, such as a second limit switch in case the first fails, or a physical barrier which causes the engine to strain and blow a fuse, thereby terminating electrical supply to the motor (Kvien and Branch 1988).

There are several alternatives for constructing the actual drive mechanisms. The simplest is an onboard motor that directly engages a drive wheel. This system works well with small shelters, where the shelter roof and motor are relatively light. The drive wheel must be adapted to provide sufficient traction, and the electrical cable supplying the motor must travel the entire length of the shelter. Another system that may use an onboard motor for larger shelters is a rack-and-pinion system in which the rack is welded to the rail. The pinion is mounted on the motor drive shaft that directly engages the rack. This system is the most expensive but provides the most traction for moving a shelter (Foale et al. 1986). A looped cable system or a sprocket-and-chain system can be run from a stationary motor, which has the advantage of solid electrical connections that do not require a wire take-up system. The looped cable works by connecting a continuous cable from one end of the tracks to the other with the shelter attached to the cable. The cable is then looped around a drive wheel attached to the motor. The two turns around the drive wheel provide sufficient traction to keep the cable taut and to move the shelter. The sprocket-and-chain system uses a looped chain attached to drive sprockets, similar to the cable system. This system provides greater positive traction than the cable system (Foale et al. 1986). On large shelters, drive mechanisms are generally attached to both sides of the shelter to avoid the twisting motion from one drive. On small, light shelters, only one side needs to be attached to the drive system.

Triggering Mechanism

The triggering mechanism for turning the shelters on and off can be tied to almost any environmental variable. Precipitation is the usual event that triggers the shelters, but wind and temperature can also be used (Ries and Zachmeier 1985). Basically, any sensor that supplies electrical voltage can be used to trigger a shelter. Regardless of which environmental cue is used, shelters also should be equipped with a manual override, so the shelter can be moved for sampling or positioned over the plots when major storms are predicted. Another key feature is inclusion of a delay switch in the circuitry, so the shelter remains in its new position for a prespecified amount of time. This prevents the shelter from constantly moving back and forth during intermittent showers. On some shelters, the move back to the open position may be delayed at night to prevent multiple moves.

Rain detectors are the most common triggering mechanism. These small sensors detect raindrops and supply a low voltage to a relay switch, which in turn triggers the shelter motor. These sensors may be triggered by as little as a single raindrop, so total exclusion of rainfall is possible. To prevent the shelter from moving back and forth, a 2- to 3-minute delay after the last raindrop is built into the sensor. A longer delay time can be added by incorporating a time-delay relay switch in the rainout shelter controls. In addition, the sensors are usually heated to eliminate early morning condensation, which would trigger the shelter.

The use of a data logger or a personal computer allows greater flexibility in managing the rainout shelter. A tipping bucket rain gauge can be programmed to allow a specified amount of rain to fall before the shelter moves, rather than moving as soon as rain is detected (Upchurch et al. 1983). The delay time can be programmed through a data logger for almost any time interval after the last detected rainfall. The data logger can also be equipped with a modem, so that the shelter can report problem conditions to a central location or be triggered from a remote location.

Belowground Water Management

Most rainout shelter designs incorporate surface and subsurface water management devices to limit lateral water movement onto the plots. Modifying precipitation seasonality or intensity can have limited impact if lateral movement is not also considered (Clawson et al. 1986; Kvien and Branch 1988). Surface water movement can be controlled on sites with a slight slope using soil berms or borders to prevent both runoff and run-on. On sites with a greater slope, run-on can be limited by berming the uphill sides of the plot. The downhill side should be left open to allow natural runoff from the precipitation treatment and to avoid ponding.

Managing subsurface water movement requires isolating the soil from surrounding treatments or control areas. In small rainout shelters, the plots are simply trenched to a 1 m depth and lined with 6-mil plastic sheeting (Weltzin and McPherson 2000; Owens, unpub. data.), although the trenching may be considerably deeper when studying mature trees (e.g., 2.5 m depth for *Prosopis glandulosa,* Jacoby et al. 1988). The plastic sheeting prevents lateral water movement, and trenching breaks root connections between the treatment area and the surrounding soil. The relatively short life of the plastic in sunlight is a disadvantage: the upper portion of the plastic, normally above the soil surface, must be covered to prevent photodegradation. A more permanent alternative is to insert corrugated metal sheeting either in a trench (Fay et al. 2000) or directly into the soil (Clark and Reddell 1990). When large rainout shelters are used, the footings and foundation for the rail system can be extended to form a water barrier. Cement foundation and footings, such as those 1.5 m deep and 10 cm thick used by Ries and Zachmeier (1985), also constrain subsurface water movement. In hot climates, this approach may not work, because the cement wall heats during the day and radiates the heat into the plot at night. The cement should be insulated from the plots to prevent altering the soil microclimate. Larsen et al. (1993) trenched around and between each individual plot in a Montana shortgrass community to a depth of 2.1 m and filled the trenches with expanding foam insulation. The foam isolated each plot from water movement and temperature gradients. The foam was covered by wooden caps to protect the foam from mechanical damage and photodegradation.

Irrigation Systems

Techniques and rationales for adding precipitation back to the plots under rainout shelters seem to be as varied as designs for the shelters themselves. Precipitation has been removed or added back to change the frequency (Fay et al. 2000), acidity (Lewin and Evans 1986), and seasonality of precipitation (Reynolds et al. 1999; Svejcar et al. 1999; Weltzin and McPherson 2000), to test for community recovery following drought (Heitschmidt et al. 1999), to select genotypes based on drought tolerance (Johnson et al. 1990; Johnson and Rumbaugh 1995), and to test for crop water use efficiency. On fixed shelters, the irrigation system is usually attached to the superstructure supporting the roof over the plots, but small plots can also be hand-watered (e.g., Weltzin and McPherson 2000). Using quarter-, half-, and full-circle spray nozzles, it is possible to irrigate subplots under a rainout shelter. If water is being applied from a pressurized system, such as a well or municipal supply, then pressure regulators and flow meters can be used to determine the application rate and amount. On the other hand, if water is being applied from an unpressurized system, such as rainfall collected off the shelters, then individual rain gauges should be used in each plot to verify the amount of water applied.

Irrigation in movable shelters can be accomplished by laying conventional irrigation pipe in the plots and applying water according to the treatments, or by designing an integral irrigation system built into the shelter that applies water from overhead (Ries and Zachmeier 1985; Nesmith et al. 1990). The advantages of an integral system are that the sides of the shelter minimize wind, so the irrigation is highly controlled, and there are no pipes or risers in the plots that interfere with normal cultural practices. Nesmith et al. (1990) designed an irrigation system for use on 48 plots (4.6 m by 6.2 m) under a single rainout shelter. The system was capable of applying rainfall up to 25 mm per hour on a plot, while only 3 mm per hour fell 0.7 m away from the plot edge. This precision allowed them to develop multiple water treatments under a single movable rainout shelter. Lewin and Evans (1986) described a sophisticated irrigation system that filtered, deionized, checked acidity, added the necessary acid for the treatment, mixed the solution, and applied it to multiple plots.

Johnson and Rumbaugh (1995) and Johnson et al. (1990) constructed line-source irrigation systems beneath rainout shelters. Simulated precipitation decreased from about 76 to 5 centimeters with increased distance from the center of the shelter. The gradient in precipitation was used to determine drought tolerance in native and introduced forage plants.

Svejcar et al. (1999) documented an interesting side effect of simulating precipitation: enhanced spring precipitation concentrated rodent use under that treatment, because the vegetation was green while surrounding vegetation was dormant. Herbivory was greatest on the spring-irrigated plots, which potentially confounded the experiment. Similarly, we irrigated plots in semiarid rangelands of southern Texas, and irrigated plots were the only green vegetation in the area during a prolonged dry period. Although the plots were fenced to prevent livestock grazing, white-tailed deer (Odocoileus virginianus) began concentrating in the area. Plots were fenced to prevent deer access, and cottontail rabbits (Sylvilagus floridanus) began concentrating in the irrigated area. The area was fenced with fine-mesh poultry netting to eliminate rabbit grazing, and Rio Grande wild turkeys (Meleagris gallopavo) flew into the plots and began grazing. By this time, there was very little vegetation left and the plots were abandoned. Therefore, depending on objectives, it may be necessary to add a substantial amount of fencing around rainout shelters to prevent herbivory.

Irrigation under rainout shelters can be fairly simple, with portable pipes, risers, and sprinklers, or fairly complex, with an integrated system of controlling fertility or acidity. Single, large rainout shelters require a precise irrigation system that will minimize drift and limit water to the appropriate plots. Small shelters can be developed so that individual irrigation treatments are under separate shelters. Irrigation in this case can be less precise because overspray is not an issue. Drip irrigation is another possibility for applying water precisely where it is needed, but canopy moisture and relative humidity may not be comparable to overhead watering. Overall, irrigation under rainout shelters is largely dependent on the arrangement of treatments and the amount of water needed.

Summary

Rainout shelters can extend our capability to study the effects of the amount and seasonality of precipitation on terrestrial ecosystems. The selection of fixed versus movable shelters largely depends on research objectives and funds available for shelter construction. Fixed shelters are less expensive to build than movable shelters but can alter the microenvironment around plants. Air and soil temperatures are likely to increase under permanent shelters, while radiation is likely to decrease. Experimental designs with fixed shelters should include either an ambient precipitation treatment both beneath and away from the shelter, or at least should monitor environmental conditions beneath and adjacent to shelters. Movable shelters can be designed to cover small or large plots. The size and weight of movable shelters dictate the type of track and motor needed. New designs for small shelters incorporate 12-volt direct-current motors for remote location use. Movable shelters can be triggered by any environmental sensor that supplies a low voltage. Small, movable shelters are particularly useful in ecological studies of heterogeneous plant communities, where an interspersion of experimental treatments is necessary.

6

The Influence of Precipitation Timing on the Sagebrush Steppe Ecosystem

TONY SVEJCAR, JON BATES,

RAYMOND ANGELL, & RICHARD MILLER

Climate influences virtually all aspects of ecosystem development and global distribution (Emmanuel et al. 1985). Some regions have experienced large climatic shifts in the recent past. The northern Great Basin of the western United States has undergone large shifts in temperature and precipitation patterns over the past 10,000 years (Morrison 1964; Mehringer and Wigand 1990; Thompson 1990). Alternating periods of cool/wet, cool/dry, warm/wet, and warm/dry conditions have caused fluctuations in composition, cover, productivity, and distribution of northern Great Basin vegetation (Wigand and Nowak 1992). Tausch et al. (1993) suggest that Great Basin climate was relatively stable during the late Tertiary period (2 to 20 million years before present), but that the Quaternary (the past 2 million years) was a period of high climatic variability. They conclude that climatic variability has resulted in plant communities that are far less stable than we previously assumed. From their assessment of paleobiological research, Graham and Grimm (1990) suggest that past climate change has resulted in individualistic changes in species distributions, rather than shifts in community boundaries. Their conclusions tend to support those of Tausch et al. (1993), that communities may not be stable in the force of climatic shifts.

Changes in seasonal climatic patterns can have a major impact on the dynamics of plant communities. In arid ecosystems there is a strong interaction between rainfall and temperature in determining plant abundance and composition. Rainfall during the hot season results in lower plant available moisture than an identical rainfall event during the cool season of the year. In arid ecosystems, even small changes in a plant's available moisture can produce major effects on plant composition. In ecosystems dominated by annuals, interannual variation in floristic composition is influenced by rainfall timing, as shown by Peco and Espigares (1994), who concluded that the timing of autumn rainfall determined the floristic composition of annual Mediterranean pastures. The yearly compositional changes were a result of germination characteristics of individual species. In an Australian pasture study, Austin et al. (1981) determined that seasonality of rainfall had more impact on pasture plant dynamics than did grazing intensity. The Intergovernmental Panel on Climate Change (IPCC) has predicted that changes in seasonal patterns of rainfall and temperature will have more impact on plant production than will changes in annual rainfall totals for large areas of Africa and North America (Ojima et al. 1993).

The impacts of climate change are of interest from a scientific standpoint, but also pose questions for management of agricultural and natural ecosystems. Hanson et al. (1993) used three general circulation models (GCMs) and a rangeland model (SPUR) to simulate outputs of a range/livestock system under different climate scenarios. They discovered that changes in production were more closely related to changes in temperature and precipitation than to changes in atmospheric carbon dioxide (CO_2). Their results were dependent on the particular GCM used in the simulation, but one pattern that emerged was an increase in rangeland production, a decrease in forage quality, and a higher year-to-year variability in production relative to current conditions. As both Hanson et al. (1993) and Helms et al. (1996) point out, farmers and ranchers have many management options and will likely adapt to changes, especially if changes occur over relatively long time frames.

The experimental approaches for investigating climate change effects on vegetation consist of indirect and direct methodologies. Indirect methods include paleobotanical studies to assess vegetation changes associated with past climatic shifts (Graham and Grimm 1990; Tausch et al. 1993; Miller and Wigand 1994; Nowak et al. 1994), comparison of vegetation patterns among regions with different climates (Cook and Irwin 1992), and correlating long-term vegetation measurements to yearly weather patterns (Passey 1982).

Although indirect methods of research are useful for predicting climate change responses and interpreting vegetation shifts, they possess a number of limitations. Paleobotanical studies are rather coarse in nature, able to describe changes in abundance of major species or functional groups but unable to detect changes for most individual species. In addition, the variation in climate over several million years can be dramatic and may not be relevant to predicting changes over the next 50 to 100 years. As with the paleobotany research, it is difficult to know whether regional climate comparisons are relevant to predicting vegetation responses. Vegetation patterns in various regions developed over evolutionary time and may or may not be indicative of what will occur at a decadal time scale. Correlation between vegetation and yearly weather patterns may not be a good indicator of what we would find with a sustained shift in climate because year-to-year fluctuations often fall within the normal variation of the ecosystem.

Direct experimental approaches have been used extensively to evaluate the impacts of elevated atmospheric CO_2 on plant growth (Amthor 1995). Most elevated CO_2 studies have been single-species experiments and thus have not addressed effects to intact multispecies ecosystems (Díaz 1995). It may not be realistic to scale up from results obtained from isolated plants growing under controlled conditions (Körner 1995). More work is needed on intact ecosystems, because interactions among plants are so important and can cause unpredicted results. In a recent study, Harte et al. (1995) used infrared radiators to warm montane meadow plots in the field. The authors concluded that vegetation will play a prominent role in determining soil microclimate response to increased temperature, a fact often obscured by climatic models. The results of Harte et al. (1995) tend to support the assertions of Körner (1995), that research from intact ecosystems may give results that would not have been predicted from controlled studies or models.

An aspect of climate change that is difficult to predict is precipitation, both amount and seasonal distribution. Research on precipitation effects (especially timing) can be difficult to conduct, which partially explains the dearth of data on this subject. As mentioned previously, precipitation distribution in the Great Basin has changed in the past, and there is reason to believe there will be shifts in the future that will affect vegetation composition, distribution, and productivity. Climate models suggest that the Great Basin may experience more summer and less winter precipitation in the future (Neilson et al. 1989). We established a study to investigate the effects of altered timing of precipitation on vegetation (annual amount held constant) in the northern Great Basin. Treatments consisted of higher summer/lower winter precipitation (spring), higher winter/lower spring precipitation (winter), and a treatment conforming to long-term precipitation distribution averages (current).

Based on regional comparisons made by Cook and Irwin (1992), we hypothesized: (1) a shift to more summer precipitation (spring treatment) would favor graminoid species over shrubs, (2) a shift to a higher percentage of winter precipitation would favor shrubs and winter annual species, and (3) the treatment receiving the average distribution would show no change relative to ambient plots.

Methods

Study Area and Experimental Design

The study was conducted on the Northern Great Basin Experimental Range (119° 43′W, 43° 29′N) in southeastern Oregon, 67 km west of Burns, Oregon. The Experimental Range is characterized by shrub steppe vegetation represented by sagebrush/bunchgrass and western juniper plant communities.

The study site is codominated by Wyoming big sagebrush *(Artemisia tridentata* subsp. *wyomingensis)* and the cold-season perennial bunchgrass species Thurber's needlegrass *(Stipa thurberiana)*, bluebunch wheatgrass *(Pseudorogenria spicata)*, and Sandberg's bluegrass *(Poa sandbergii)*. Elevation at the site is 1380 m and the ground is level (0 to 1 percent slope). Soils are well drained and underlain by a duripan between 40–50 cm. Soils on the site were classed as a Vil-Decantl Variant-Ratto complex (Lentz and Simonson 1986). Field capacity of soils is 23 percent (0–15 cm) and 25 percent (15–30 cm) for gravimetric soil water content. Climate is continental with cold-wet winters and dry-warm summers. Annual precipitation at the Experimental Range has averaged 300 mm since measurements began in the 1930s. Distribution of precipitation during this period was 60 percent from October to March, 30 percent from April to June, and 10 percent from July to September. It is important to note that annual precipitation in the Great Basin is extremely variable from year to year. For example, at the Experimental Range the wettest years on record were 1938 and 1993, each with about 530 mm of precipitation. The driest year on record was 1994 with only 140 mm of precipitation.

To assess the effects of timing on soil water and plant community dynamics, five rainout shelters were constructed in 1994. The design of the fixed location rainout shelters and associated irrigation system is described in Svejcar et al. (1999). Rainout shelters were 30 by 12 m in size. Shelters are open on the sides and until 1998 were covered with transparent fiberglass. The fiberglass was replaced in the summer of 1998 with Dynaglass®, a clear polycarbonate material.[1] Precipitation under the shelters was applied by an overhead sprinkler system.

Precipitation treatments under each shelter were designated as "winter," "spring," and "current." Treatment plots were 10 m by 12 m in size and included a 2 m buffer strip on all sides. The winter distribution treatment received the majority of precipitation (80 percent) between October and March; the spring distribution treatment received the majority of its precipitation (80 percent) between April and July; the current distribution treatment received precipitation conforming to long-term (50 years) distribution patterns at the Experimental Range. The target precipitation distribution schedules for each treatment are shown in table 6.1. The target was for all shelter treatments to receive a total of 203 mm of water annually. During the initial year of study, we found that application of 300 mm of water (the long-term average annual precipitation) resulted in surface puddling and saturated soil. Soil moisture was much higher than would have been expected based on historical precipitation and soil moisture data, because our method of application was more effective at increasing soil moisture than a comparable amount of natural precipitation. This region is characterized by low intensity, relatively long duration storms, with infrequent thunderstorms. The discrepancy between natural precipitation and sprinkler application is probably a result of duration and intensity of moisture fall. Therefore, we chose to decrease the total amount of precipitation applied to the shelter treatments to about 200 mm.

Ambient treatment plots of identical size were located south of each shelter. Ambient precipitation was measured using a tipping bucket rain gauge, and amounts were automatically recorded using an electronic data logger. Ambient plots received natural precipitation; thus, amounts and patterns varied by year.

The experimental design was a randomized complete block with four treatments replicated five times. Understory biomass and cover were compared between treatments (among and by year) using General Linear Model (GLM) statistical techniques for a randomized block design. Main effects for understory biomass and cover were year and treatment. Soil water content was analyzed using a repeated measures analysis of variance (ANOVA) for a randomized block design. Main effects for soil water content were treatment, soil depth, and time. Data was tested for normality (SAS Institute 1988); data not normally distributed were log-transformed to stabilize variance. When interactions were significant, means were separated using Fisher's protected

TABLE 6.1.

Precipitation (mm) patterns for the ambient treatment and the shelter treatments (Current, Winter, Spring) in 1997–98 and 1998–99. Values are the proportion of total precipitation received and the total annual precipitation amounts during the course of the water year (October–September).

	TREATMENT PERIODS			
Year/Treatment	Winter (October–April)	Spring–Summer (May–July)	Fall (August–September)	Precipitation Total
Target application				
Ambient[1]	180	90	30	300
Current	153	40	10	203
Winter	183	20	0	203
Spring	45	158	0	203
Precipitation applied 1997–98				
Ambient[2]	161	108	25	294
Current	122	74	13	209
Winter	182	22	0	204
Spring	55	152	0	207
Precipitation applied 1998–99				
Ambient[2]	115	24	10	149
Current	133	60	12	205
Winter	185	21	0	205
Spring	48	156	0	204

[1]Long-term average at the Experiment Range.

[2]Ambient precipitation levels in 1997–98 and 1998–99 underestimate actual amounts, because the rain gauge was not 100 percent efficient at capturing snow.

least significant difference (LSD) procedure. Statistical significance of all tests was assumed at $P < 0.05$.

Vegetation Measurements

Herbaceous biomass was estimated in September 1998 and June 1999. In each treatment replicate, five 1.0 m² quadrants were clipped for herbaceous biomass. Biomass was separated into perennial and annual components.

Herbaceous and total ground cover were visually estimated in 1994, 1998, and 1999 within 0.2 m² frames. Frames were placed every meter along an 8 m transect line. Herbaceous cover was separated into perennial and annual components.

Plant phenology was monitored on a weekly to biweekly schedule, from growth initiation to seed scatter over five growing seasons (1995–1999). Plants monitored were Wyoming big sagebrush, Thurber's needlegrass, squirreltail *(Sitanion hystrix)*, blue-eyed Mary *(Collinsia parviflora)*, and as a group we monitored the perennial forbs pale agoseris *(Agoseris glauca)*, western hawksbeard *(Crepis occidentalis)*, and tapertip hawksbeard *(C. acuminata)*. Three plants of each species were monitored in each treatment replicate (15 subsamples per treatment).

Precipitation Application, Soil Water Content, and Temperature

Water applied to shelter treatments was measured with five rain gauges permanently placed in each replicate. Rain gauges were constructed using 2 L plastic soft drink containers and were anchored to the ground with steel rods (Wrage et al. 1994). Measurements were done immediately after water was applied.

Soil water content was determined gravimetrically in 1998 and 1999. Soil water measurements were collected at 0–15 cm and 15–30 cm intervals every two weeks during the growing season (April–September). Two randomly placed subsamples were collected for each depth in each treatment replicate. Soils were weighed, dried at 106°C for 48 hours, and reweighed to determine gravimetric water content.

Soil temperature was recorded in 1996, 1997, and 1998 in each treatment plot with thermocouples placed 5 cm below the surface. Concurrent with temperature, soil surface wetness was estimated at two locations in each plot using a granular matrix sensor (Watermark, Irrometer Co., Riverside, CA) buried at 5 cm. Average hourly temperature and moisture data were estimated from measurements taken at 5 min intervals using onsite data loggers. Data stored in the data loggers was downloaded weekly to a computer. Average monthly soil temperatures and annual soil moisture were calculated from the hourly data. Soil moisture (percent) was estimated from the sensor resistances using a regression equation developed from gravimetric soil water measurements. Although the matrix sensors are not quantitatively accurate when soils dry below about -300 kPa, they were appropriate for this study because we were primarily interested in relative differences in surface wetness as affected by different watering treatments.

TABLE 6.2.
Mean (± 1 SE) total biomass (kg/ha) and biomass percentages of perennial and annual vegetation for the four treatments (n = 25).

Year	TREATMENT			
	Winter	Spring	Current	Ambient
September 1998				
Biomass (kg/ha)	338 ± 56 b[1]	201 ± 40 a	281 ± 46 a	371 ± 67 b
% Perennial	85.1 ± 6.0 c	99.4 ± 0.3 d	98.7 ± 0.6 d	96.9 ± 0.8 d
% Annual	14.9 ± 6.0 e	0.6 ± 0.3 f	1.3 ± 0.6 f	3.1 ± 0.8 f
June 1999				
Biomass (kg/ha)	472 ± 87 h	148 ± 30 g	464 ± 66 h	426 ± 47 h
% Perennial	81.9 ± 5.6 m	95.6 ± 2.5 o	85.9 ± 4.3 m	99.4 ± 0.3 o
% Annual	18.1 ± 5.6 z	4.4 ± 2.5 y	14.1 ± 4.3 z	0.6 ± 0.3 y

[1]Within a row, means followed by the same lowercase letter were not different ($P > 0.05$).

Results

Plant Community Dynamics

BIOMASS.   Total biomass production in September 1998 and June 1999 was significantly less in the spring treatment versus other treatments (table 6.2). As a percentage of total biomass production, the winter treatment had the greatest amount of annual plant production. Annual plant production in the winter treatment was significantly greater than spring and ambient treatments in both years. In 1999, the percentages for annual and perennial production were not significantly different between current and winter treatments.

COVER.   Prior to shelter construction and treatment initiation in 1994, there were no differences in herbaceous (perennial, annual) cover among experimental units (table 6.3). Bare ground was somewhat lower in the ambient and spring treatments compared with current and winter treatments in 1994 but did not differ statistically.

In 1998 and 1999, herbaceous cover increased in all treatments from 1994. The increases in cover were not the same among treatments. Total herbaceous cover and annual cover were significantly greater in the winter versus the other treatments. The current treatment also had significant gains

TABLE 6.3.
Mean ($\pm$ 1 SE) total herbaceous cover and perennial and annual vegetation cover for the four treatments, given as a percentage of total area in treatment plot (n = 15).

Year/Response Variable	TREATMENT			
	Winter	Spring	Current	Ambient
1994				
Bare ground	71.8 ± 2.8	64.2 ± 3.0	66.8 ± 2.5	60.2 ± 4.7
Herbaceous cover	9.1 ± 0.8	9.9 ± 0.7	9.8 ± 0.7	9.7 ± 0.9
Perennial	8.4 ± 0.7	9.8 ± 0.7	9.6 ± 0.7	9.6 ± 1.0
Annual	0.7 ± 0.3	0.1 ± 0.0	0.2 ± 0.0	0.1 ± 0.0
1998				
Bare ground	38.2 ± 3.2 a[1]	58.8 ± 3.0 c	45.0 ± 3.9 b	68.3 ± 3.7 d
Herbaceous cover	28.1 ± 2.0 o	15.4 ± 2.5 m	19.8 ± 1.6 t	17.4 ± 1.0 mt
Perennial	23.4 ± 1.9 s	15.2 ± 2.4 q	19.0 ± 1.5 r	16.4 ± 1.0 rq
Annual	4.7 ± 1.4 y	0.3 ± 0.2 x	0.8 ± 0.3 x	1.0 ± 0.2 x
1999				
Bare ground	35.0 ± 2.3 a	55.6 ± 2.5 c	43.2 ± 2.4 b	59.9 ± 3.8 c
Herbaceous cover	38.0 ± 1.8 p	12.8 ± 2.4 m	30.5 ± 2.3 o	22.0 ± 0.6 t
Perennial	29.6 ± 2.2 s	12.5 ± 2.3 q	27.3 ± 2.1 s	21.3 ± 1.0 r
Annual	8.4 ± 2.3 y	0.3 ± 0.1 x	3.2 ± 1.1 x	0.6 ± 0.1 x

[1] Within a row, means followed by the same lowercase letter were not different ($P > 0.05$).

in cover, though not as great as those in the winter treatment. Perennial plant cover was significantly lower in the spring treatment versus all other treatments in 1999. Bare ground was highest in the ambient treatment versus the shelter treatments in 1998.

REPRODUCTIVE DEVELOPMENT.   Phenological development of all the species monitored was affected by the precipitation treatments (table 6.4). Sagebrush reproductive success (1995–1999) was highly variable in the ambient treatment (47–100 percent) in contrast to all shelter treatments, which had more consistent reproductive development (80–100 percent). Thurber's needlegrass and squirreltail reproductive success was significantly lower in the spring treatment versus the other treatments in 1995, 1998, and 1999. Perennial forb reproductive success was highly variable for all treatments during the study but was consistently the lowest in the spring treatment. Reproductive success for the annual forb *Collinsia* was significantly lower in the spring versus other treatments in all years.

TABLE 6.4.
Mean ($\pm$ 1 SE) growth development success percent by treatment and year for *Artemisia tridentata* subsp. *wyomingensis*, *Stipa thurberiana*, *Sitanion hystrix*, all perennial forbs, and *Collinsia parviflora* (n = 15). Growth development success was defined as the percentage of observed plants that completed all growth stages between growth initiation and reproduction.

Species/Year	TREATMENT			
	Winter	Spring	Current	Ambient
A. tridentata				
1995	100.0 $\pm$ 0.0	100.0 $\pm$ 0.0	100.0 $\pm$ 0.0	100.0 $\pm$ 0.0
1996	86.7 $\pm$ 8.2	86.7 $\pm$ 8.2	86.7 $\pm$ 8.2	66.7 $\pm$ 18.3
1997	93.3 $\pm$ 6.7	93.3 $\pm$ 6.7	100.0 $\pm$ 0.0	93.3 $\pm$ 6.7
1998	100.0 $\pm$ 0.0	100.0 $\pm$ 0.0	100.0 $\pm$ 0.0	100.0 $\pm$ 0.0
1999	80.0 $\pm$ 8.2	100.0 $\pm$ 0.0	80.0 $\pm$ 8.2	46.7 $\pm$ 17.0
Average	92.0 $\pm$ 2.9b[1]	96.0 $\pm$ 2.2 b	93.3 $\pm$ 2.7 b	81.3 $\pm$ 6.4 a
S. thurberiana				
1995	100.0 $\pm$ 0.0	40.0 $\pm$ 6.7	100.0 $\pm$ 0.0	100.0 $\pm$ 0.0
1996	100.0 $\pm$ 0.0	93.3 $\pm$ 6.7	100.0 $\pm$ 0.0	60.0 $\pm$ 12.5
1997	100.0 $\pm$ 0.0	73.3 $\pm$ 12.5	86.7 $\pm$ 8.2	86.7 $\pm$ 8.2
1998	100.0 $\pm$ 0.0	66.7 $\pm$ 10.5	100.0 $\pm$ 0.0	100.0 $\pm$ 0.0
1999	90.0 $\pm$ 6.1	50.0 $\pm$ 7.9	95.0 $\pm$ 5.0	95.0 $\pm$ 5.0
Average	98.0 $\pm$ 1.4 e	64.7 $\pm$ 5.3 d	96.3 $\pm$ 2.0 e	88.3 $\pm$ 6.4 e
S. hystrix				
1995	100.0 $\pm$ 0.0	80.0 $\pm$ 13.3	100.0 $\pm$ 0.0	100.0 $\pm$ 0.0
1996	100.0 $\pm$ 0.0	86.7 $\pm$ 8.2	100.0 $\pm$ 0.0	93.3 $\pm$ 6.7
1997	100.0 $\pm$ 0.0	86.7 $\pm$ 8.2	100.0 $\pm$ 0.0	93.3 $\pm$ 6.7
1998	100.0 $\pm$ 0.0	66.7 $\pm$ 18.3	100.0 $\pm$ 0.0	100.0 $\pm$ 0.0
1999	100.0 $\pm$ 0.0	60.0 $\pm$ 12.7	95.0 $\pm$ 5.0	95.0 $\pm$ 5.0
Average	100.0 $\pm$ 0.0 h	73.3 $\pm$ 5.8 g	99.0 $\pm$ 1.0 h	96.7 $\pm$ 2.0 h
Perennial forb				
1995	20.0 $\pm$ 8.2	6.7 $\pm$ 6.7	53.3 $\pm$ 17.0	73.3 $\pm$ 12.5
1996	33.3 $\pm$ 10.5	20.0 $\pm$ 8.2	26.7 $\pm$ 12.5	46.7 $\pm$ 13.3
1997	73.3 $\pm$ 16.3	0.0 $\pm$ 0.0	80.0 $\pm$ 13.3	66.7 $\pm$ 10.5
1998	73.3 $\pm$ 6.7	20.0 $\pm$ 13.3	93.3 $\pm$ 6.7	33.3 $\pm$ 14.9
Average	50.0 $\pm$ 7.5 n	11.7 $\pm$ 4.4 m	63.3 $\pm$ 8.2 n	55.0 $\pm$ 7.0 n
C. parviflora				
1995	100.0 $\pm$ 0.0	26.7 $\pm$ 12.5	100.0 $\pm$ 0.0	100.0 $\pm$ 0.0
1996	100.0 $\pm$ 0.0	73.3 $\pm$ 12.5	100.0 $\pm$ 0.0	100.0 $\pm$ 0.0
1997	100.0 $\pm$ 0.0	0.0 $\pm$ 0.0	100.0 $\pm$ 0.0	100.0 $\pm$ 0.0
1998	100.0 $\pm$ 0.0	6.7 $\pm$ 6.7	100.0 $\pm$ 0.0	100.0 $\pm$ 0.0
Average	100.0 $\pm$ 0.0 y	26.7 $\pm$ 7.9 x	100.0 $\pm$ 0.0 y	100.0 $\pm$ 0.0 y

[1] Within a row, means followed by the same lowercase letter were not different ($P > 0.05$).

Precipitation Application and Soil Water Content

Water applied to the shelter treatments and recorded for the ambient plots in 1997–98 and 1998–99 is shown in table 6.1. Water applied to the shelter treatments does not conform exactly to the target schedule. This is especially true for February, when air temperatures were too cold to apply water. Watering during cold winter months was limited because of frequent water line and sprinkler system breakage and because water had a tendency to freeze on plants unless air temperatures were well above freezing. Ambient precipitation recorded at the shelters probably underestimated actual precipitation because the tipping bucket rain gauge used did not capture all moisture received as snow.

Precipitation patterns in the ambient treatment illustrate that variability among years is high in this system. The ambient treatment received almost three times as much moisture in the spring-summer period in 1997–98 compared with the same period in 1998–99. Ambient annual precipitation in 1998–99 was also about 40 percent less than in 1997–98.

Soil water content differed among treatments in the 1998 and 1999 growing seasons (fig. 6.1). In 1998, ambient soil water content was greater than the other treatments between mid-April and early June. Spring soil water content was greater than the other treatments between late June and mid-August. In 1999 soil water in the spring treatment was less than all other treatments until late May. Soil water content was greater in the spring treatment versus the other treatments between late June and mid-August.

Soils at both depths in the spring treatment never approached field capacity (24 percent gravimetric soil water) nor became thoroughly wetted through the profile, despite receiving the same amount of water as the current and winter treatments (table 6.1). Soils in all other treatments started the growing seasons above or just below field capacity in 1998 and 1999.

Surface soil moisture clearly reflected the shifts in precipitation distribution for each treatment, with the spring treatment being significantly drier during the winter period, and wetter than current and winter during the April–June period (fig. 6.2). During the three-year period, current and winter plots tended to dry more quickly than the ambient plots, perhaps because actual precipitation was at or above average during this time.

Soil temperature tended to be lowest in the ambient treatment (fig. 6.3). These differences were probably caused by the insulating effect of the shelters, even though they were open on all sides and well ventilated. The effect seemed to be most pronounced in summer, when soils were dry and plant

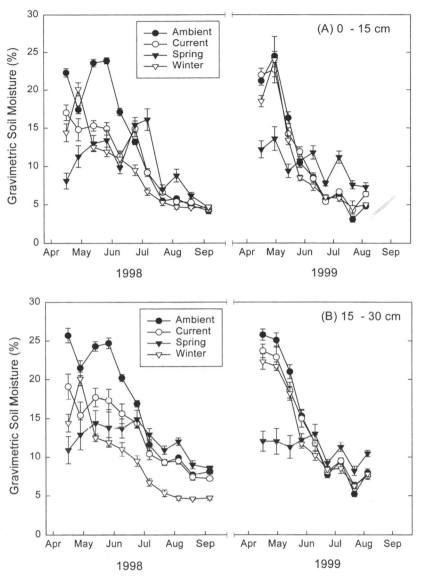

Figure 6.1. Soil moisture (percent) at 0–15 cm (A) and 15–30 cm (B) for the shelter treatments and ambient plots during the growing seasons of 1998 and 1999. Vertical bars are one standard error of the means.

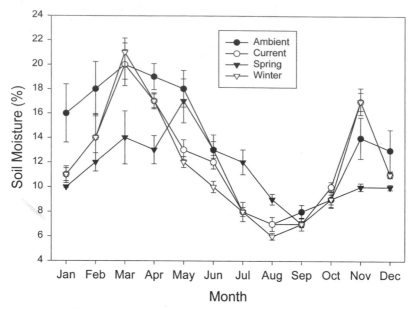

Figure 6.2. Monthly average soil moisture at 5 cm for shelter treatments and ambient plots. Values were calculated from hourly data measured with Watermark® soil sensors. Vertical bars are one standard error of the mean.

growth was decreasing. During the April to June period, soil temperatures on all treatments were similar.

Discussion

Vegetation Response

Precipitation timing affected the growth and structure of the sagebrush steppe community we studied. The effects differed from our initial expectations. Based on ecosystem comparisons such as that of Cook and Irvin (1992) and other studies of grasslands and shrublands (e.g., Coupland 1979; Sala et al. 1989), we predicted that the winter treatment would favor shrubs and the spring treatment would favor grasses. Precipitation during the dormant season (winter) should recharge the lower part of the soil profile, and thus favor tap-rooted species, whereas growing season (spring) precipitation should favor the fibrous-rooted grasses that are effective at using moisture from the upper levels of the soil profile (Coupland 1979; Yoder et al. 1998). Our results contradict those initial predictions. Standing herbaceous biomass was

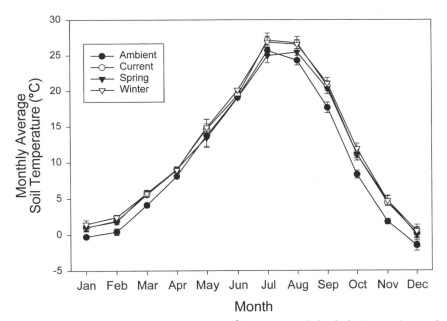

Figure 6.3. Monthly average soil temperature (°C) at 5 cm soil depth during 1996 to 1998. Vertical bars are one standard error of the mean.

consistently lowest in spring compared with other treatments (table 6.2). The spring treatment also had the largest percentage of bare ground and the lowest percentage of herbaceous cover when compared with other shelter treatments (table 6.3). Thus plant productivity and cover may have been limited by the inability of many species to fully complete their life cycle in the spring treatment (table 6.4).

There are two potential explanations for the low productivity of the spring treatment: (1) soil moisture was inadequate during a critical growth period, or (2) the later application of precipitation in the spring treatment resulted in lower plant available moisture, as soil moisture never approached levels developed in the other treatments (fig. 6.1). The fact that the annual forb *Collinsia* did not complete its life cycle (and rarely emerged) during any year of the study in the spring treatment suggests that moisture was lacking during a critical period. In all the other treatments in all years, *Collinsia* completed its life cycle 100 percent of the time (table 6.4). These results demonstrate the utility a species can have as an indicator of a climatic scenario.

The winter treatment was more conducive to growth of annual plants than was the spring treatment (tables 6.2 and 6.3). Production and cover of annuals was somewhat variable in the current and ambient treatments. The production of annuals in the ambient treatment can probably be explained by the distribution of natural rainfall. During the 1998–99 season, there was very little natural rainfall from March through May, and annual plant production was low in the ambient plots. Most of the annuals in the community we studied were winter annuals (Cronquist et al. 1977), and late winter/early spring moisture appears critical for their development. The higher cover of annuals in the winter treatment was consistent with our original hypothesis.

The dominant shrub, Wyoming big sagebrush, was not influenced significantly by any of the shelter treatments (table 6.4). There was a tendency for sagebrush to reach more advanced stages of phenology in all shelter treatments compared with the ambient plots. The slightly higher temperatures under the shelters or the manner in which water was applied might provide an explanation. Sagebrush overwinters a portion of its leaves and can be photosynthetically active during the winter (Caldwell 1979), whereas the other species we studied initiate growth in the spring. Sagebrush would, therefore, be more likely to be influenced by the higher early season temperatures (fig. 6.3) than would the other species. The slightly lower values for sagebrush development in the winter treatment compared with current and spring (table 6.4) may be because this species often responds to summer rains with increased reproductive shoot development (Evans et al. 1991). The difference is not significant at this point, but it bears watching in the future. Canopy cover estimates also do not show any significant differences among treatments for sagebrush (Bates et al. 1999). It appears that sagebrush is less likely to be influenced in the short term by climatic shifts relative to many of the herbaceous species.

Shelter Effects

The effects of the rain shelters on microclimate were discussed in a technical note published previously (Svejcar et al. 1999). We can add a little more detail to this discussion from data presented in this chapter. Average soil temperature is warmer under the shelters than in ambient plots (see fig. 6.3 for a seasonal comparison of soil temperatures). The differences under the shelters and in ambient plots are evident during all seasons except spring. The greatest similarity among treatments occurred in April, May, and June, which is also the period of maximum plant growth. Soil temperatures for different treatments under the shelters were similar most of the year. The only differences occurred in July and August, when spring plots were slightly cooler than current or winter plots (fig. 6.3). This is because spring plots received precipitation later into the summer than other treatments (table 6.1), had higher July and August soil moisture (fig. 6.1), and would have experienced some degree of evaporative cooling.

The general approach employed in this study was successful; i.e., we were able to keep total precipitation constant while altering distribution (table 6.1). The only problems we encountered were with the February waterings, when on occasion it was necessary to delay water application for a week or two due to low temperatures. However, when temperatures were too low for watering, they were also too low for much plant growth. Another technical consideration was the wind. We avoided watering during windy periods, and during the spring it was sometimes necessary to apply the water at sunrise before convective winds began.

Conclusions

Changes in precipitation distribution have the potential to influence the structure and productivity of sagebrush steppe vegetation. However, our results do not conform to the original hypothesis that winter precipitation will favor shrubs, and spring/summer precipitation will favor grasses. In this study, shifting precipitation to a spring/early summer pattern had a negative effect on the plant community in terms of herbaceous productivity, vegetation cover, and the ability of some key plant species to reproduce. Herbaceous plants in the environment of the Great Basin are physiologically adapted to a winter/early spring precipitation pattern, where reliable soil water recharge occurs prior to the growing season. Development of a spring/summer precipitation pattern would result in declines and, potentially, the eventual loss of some native annual and perennial forbs. Biomass production would also be reduced. Wildlife, domestic livestock, and other organisms that depend on the production of herbaceous annual and perennial vegetation would be adversely affected by a spring moisture regime. The shelter results also suggest there would be an increase in bare ground with a spring moisture pattern. More bare ground could increase soil erosion, and the open sites created by loss of native plant species may permit invasion by noxious weeds.

We propose two possible explanations for the negative effects of shifting precipitation to spring from winter: (1) the spring/early summer distribution resulted in plant stress during a critical early growth period (probably March), or (2) applying precipitation later in the growing season reduced effective soil moisture for plant growth (because of higher evapotranspiration) compared with winter application. This study demonstrates that experimental research conducted in the field can provide an important test of assumptions drawn from observational studies.

Notes

1. The mention of trade names does not indicate an endorsement by USDA-ARS or Oregon State University.

7

Ecological Consequences of
Drought and Grazing on Grasslands
of the Northern Great Plains

RODNEY K. HEITSCHMIDT & MARSHALL R. HAFERKAMP

Managed grazing is the principal use of rangelands, and as such, the magnitude of animal products derived from rangelands is linked closely to quantity and quality of herbage produced. Likewise, grazing has profound impacts on both the structure and function of ecological systems relative to kinds and numbers of plant species present and rates of energy flow and nutrient cycling (e.g., see Heitschmidt and Stuth 1991; Vavra et al. 1994; Eldridge and Fruedenberger 1999). Thus, any factor affecting quantity and/or quality of herbage produced over either time or space has both economic and ecological implications.

Water is the principal abiotic variable affecting the productivity and distribution of rangeland ecosystems (Rosenzweig 1968; Webb et al. 1983; Sala et al. 1988; Stephenson 1990; Scholes 1993). Typically, annual precipitation accounts for about 90 percent of the variance in primary productivity (Le Houerou and Hoste 1977; Webb et al. 1978; Walter 1979; Foran et al. 1982; Sala et al. 1988). Because drought is common in rangeland environments, intra- and interseasonal growing conditions vary considerably. Moreover, current climate change models predict both temperature and precipitation extremes will be of greater magnitude than current extremes (Office of Science and Technology Policy 1996). Thus, greater temporal and spatial variation in the quantity and quality of herbage produced can be expected with accompanying economic and ecological impacts.

The broad objective of the research reported in this chapter was to examine the potential interactive effects of severe drought and grazing on Northern Great Plains rangelands. Specific response variables examined included aboveground net primary production and plant species composition (Heitschmidt et al. 1999), tiller (Eneboe 1996) and root growth (Hild et al. 2001) dynamics, xylem water potentials, germinable seed bank (Hild et al. 2001), and water yield and nutrient transport (Emmerich and Heitschmidt 2002).

Study Area and Treatments

Research was conducted during the 1993–1996 growing seasons at the Fort Keogh Livestock and Range Research Laboratory located near Miles City, Montana (46°22' N, 105°5' W). Regional topography ranges from rolling hills to broken badlands, with small intersecting streams that flow into large permanent rivers, which meander through broad, nearly level valleys. The potential natural vegetation on the 22,500 ha station is a grama-needlegrass-wheatgrass *(Bouteloua-Stipa-Agropyron)* mix (Küchler 1964). Long-term annual precipitation averages 34 cm with about 60 percent received during the 150-day, mid-April to mid-September growing season (fig. 7.1). Average daily temperatures range from -10°C in January to 24°C in July, with daily maximum temperatures occasionally exceeding 37°C during summer and daily minimums occasionally dipping below -40°C during winter.

Study plots were twelve 5 by 10 m nonweighing lysimeters constructed in 1992 on a gently sloping (4 percent) clayey range site (Kobase silty clay loam; fine, montmorillonitic, frigid, Aridic Ustochrepts). Lysimeters were arranged perpendicularly along a 65 m transect in two groups of six lysimeters with a 5 m area between groups. They were constructed by filling 12 cm wide by 2 m deep perimeter trenches and, aboveground, 12 cm wide by 15 cm tall wooden foundations with urethane foam insulation. Each lysimeter was equipped with two, 1 m deep, soil water monitoring access tubes, one each upslope and downslope. Likewise, each lysimeter was equipped with two, 1 m deep, minirhizotron root observation tubes, one each upslope and downslope. Each lysimeter was also equipped with a surface water runoff collection system consisting of a small (about 0.2 m²) concrete collection apron with underground plumbing for transporting water and sediment to individual fiberglass collection tanks. The study area had not been grazed by livestock since 1988.

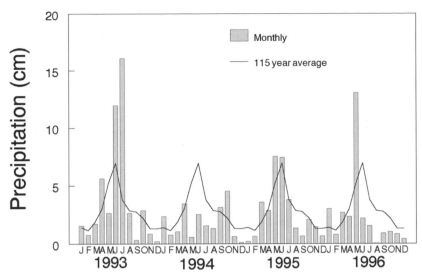

Figure 7.1. Monthly precipitation (cm) from January 1993 through December 1996 and long-term (115-year) average at Miles City, Montana. (from Heitschmidt et al. 1999)

An automated rainout shelter was constructed to control the amount of precipitation received on one of the two sets of six lysimeters (Larson et al. 1993). The 12 by 35 m metal framed roof was mounted on 15 cm diameter plastic wheels atop seven, 5 cm wide rails, which were stationed about 75 cm above the soil surface. Rails extended from top edge (i.e., upslope) to about 15 m below the bottom edge of the lysimeters. Rails were located directly over lysimeter borders. The shelter was equipped with a moisture sensitive conductance plate that, when wetted, activated a small electric motor and its associated drive system, which moved the shelter across the plots.

Twice replicated treatments were as follows: (1) graze both the year of (i.e., 1994) and the year after (i.e., 1995) simulated drought, hereafter referred to as the 94–95 grazed treatment; (2) graze only during the year of the drought, hereafter referred to as the 94 grazed treatment; and (3) no grazing either year, hereafter referred to as the ungrazed treatment. These same three treatments were repeated in the nondrought set of six lysimeters. Plots were grazed intensively with six ewes and their twin lambs for a few hours in early June and early July of both 1994 and 1995.

The simulated drought was imposed from late May to mid-October 1994. During this time, total ambient precipitation was 14.66 cm, whereas total received on the drought plots was 1.02 cm. The 1.02 cm was received during the first three weeks of the imposed drought, when timing adjustments to the automated components of the rainout shelter were being made.

Field and Laboratory Methodology

Vegetation Growth Dynamics and Aboveground Productivity

Herbage biomass was estimated monthly by clipping 10 randomly located, 250 cm² circular quadrats per lysimeter. Relative values of abundance were assigned to all species in each quadrat; however, only the most abundant species were clipped individually with minor, functionally similar species combined. Species/functional groups were western wheatgrass (*Pascopyrum smithii* Rydb. [Love]); needle-and-thread grass (*Stipa comata* Trin. & Rupr.); warm-season perennial shortgrasses, most of which was blue grama (*Bouteloua gracilis* [H.B.K.] Lag. ex Griffiths) with a small amount of buffalograss (*Buchloe dactyloides* [Nutt.] Engelm.); other warm-season perennial grasses, of which sand dropseed (*Sporobolus cryptandrus* [Torr.] A. Gray) was the dominant species; *Bromus* sp., which was principally Japanese brome (*Bromus japonicus* Thunb. ex Murr.) with a small amount of downy brome (*Bromus tectorum* L.); other cool-season perennial grasses, of which Sandberg's bluegrass (*Poa sandbergii* Vasey) was dominant; other cool-season annual grasses, of which sixweeks fescue (*Festuca octoflora* Walt.) and little barley (*Hordeum pusillum* Nutt.) were dominant; forbs; plains prickly pear cactus (*Opuntia polyacantha* Haw.); and shrubs. Herbage was dried at 60°C for a minimum of 48 hours before hand separation of live (i.e., green) and dead (i.e., brown) material and weighing.

Herbaceous aboveground net primary production was estimated by functional group (i.e., cool-season perennial grasses, cool-season annual grasses, warm-season perennial grasses, and forbs) by summing increases in live biomass. Total herbage production was estimated by summing functional group estimates. For a more detailed description of methods, see Heitschmidt et al. (1999).

Tiller Growth Dynamics

Six 14 by 18 cm quadrats were randomly located in each lysimeter in the spring of 1994. In May of both 1994 and 1995, five live tillers each of western wheatgrass and blue grama were permanently identified in each quadrat using colored wire rings. Tiller measurements were taken weekly from mid-May to mid-August with additional measurements collected in late September near the end of each growing season. Variables monitored included tiller height/length, phenological stage including senescence, and number of lamina. Plant densities were estimated in May of both 1994 and 1995. Blue grama and western wheatgrass standing biomass were estimated monthly by the harvest method, whereas tiller biomass dynamics were monitored using height-weight linear regression procedures. Tiller relative growth rates (RGR) were calculated following Radford (1967). Patterns of RGR and absolute growth rate (AGR) were examined seasonally across three time periods: (1) the last two weeks of May before the first grazing event (mid-spring); (2) the month of June between grazing events (late spring); and (3) the July–October postgrazing period (summer). Annual aboveground net primary productivity was estimated by summing positive increments in monthly biomass estimates. For a more detailed description of methods, see Eneboe (1996).

Root Growth

Root growth was monitored in situ using two minirhizotrons (Richards 1984) per lysimeter. The number of root intersects crossing circumscribed lines was recorded about every two weeks during the 1993 through 1996 growing seasons. Sample depths were 2.5 and 10 cm below the soil surface and every 10 cm increment thereafter to a depth of 90 cm. Data were summed within soil horizons and averaged by number of soil depths contained within the four horizons (i.e., A = 2.5 and 10 cm depths, Bw = 20, 30, and 40 cm depths, Bk = 50 and 60 cm depths, and C = 70, 80, and 90 cm depths). For a more detailed description of methods, see Hild et al. (2001).

Xylem Water Potential

Xylem water potentials of blue grama and western wheatgrass plants growing in the ungrazed drought and nondrought treatments were estimated on seven dates between 18 May and 28 July 1994 and six dates between 15 May and 2 August 1995. Measurements were made at 0430 and 1230 hours on the youngest fully expanded leaves of individual plants, using standard pressure chamber techniques (Scholander et al. 1965).

Seed Bank

In situ seed banks residing in soil surface litter and the top 3 cm of soil were sampled annually from 1993 through 1996. Litter and associated seeds were collected by vacuum from within six randomly located, 100 cm^2 quadrats per lysimeter. The vacuumed 100 cm^2 soil surface was then cored to a depth of

3 cm. Germinable seed banks were assessed by counting emerging seedlings in a controlled greenhouse environment of 12 hr photoperiods and diurnal temperatures of 13–21°C. Samples were incubated for 12 weeks each year with emerging seedlings counted and recorded by species on a biweekly basis. Data were summarized by functional groups (i.e., cool-season perennial grasses, warm-season perennial grasses, cool-season annual grasses, and forbs). For a more detailed description of methods, see Hild et al. (2001).

Water Yield and Nutrient Transport

For the purposes of this chapter, we define water yield as surface runoff water. Thus,

$$WY = P - DP - ET + SW \quad [\text{eq. 1}]$$

wherein, WY is water yield, P is precipitation, DP is deep percolation, ET is evapotranspiration, and SW is change in soil water. On-site precipitation was measured directly using standard rain gauges. Runoff (i.e., water yield) was measured using the collection basins. Because study site soils included a water impermeable clay layer about 1 m below the soil surface, deep percolation was assumed to be zero. Soil water was estimated at depths of 15, 30, 60, 90, and 120 cm at the beginning and end of each month using a dielectric soil water probe. The final component of the equation, evapotranspiration, was subsequently estimated by the equation

$$ET = P - WY - DP + SW \quad [\text{eq. 2}]$$

Nutrient transport was estimated by first assessing pretreatment (i.e., 1993) nutrient concentrations in surface soils and, thereafter, measuring nutrient concentrations in precipitation, surface runoff water, and soil sediment. Pretreatment soil samples consisted of eight randomly located, 4 cm deep soil samples per lysimeter. Saturated extracts were used to estimate pH, electrical conductivity, and soluble nutrient concentrations of Ca, Mg, K, Na, NO_3, NH_4, PO_4, Cl, SO_4, and HCO_3. Additional variables examined included soil texture, organic matter content, and cation exchange capacity. Precipitation samples were collected from rain gauges, and surface runoff water samples were collected from the basin collection system. Samples of sediment were collected from decanted surface runoff water samples that had been dried at 65°C and weighed to estimate total sediment production. Water soluble ion concentrations in precipitation, runoff, and sediment samples were determined using the same procedures as soil samples. Sample collection was restricted to the May through October time period when soils were not frozen and precipitation was in the form of rain. Samples were collected from each lysimeter after each rainfall event. For a more complete description of methods, see Emmerich and Heitschmidt (2002).

Data Analyses

Data were summarized by sample date, year, and grazing/drought treatment. Most statistical analyses (see exceptions below) involved using various analysis of variance models with main and interactive effects of grazing treatment, drought treatment, month, and/or year. Drought and grazing treatments were normally treated as between-plot (i.e., lysimeter) effects; thus, the error term for testing for these effects and their associated interaction was plot nested within drought and grazing treatment. When years and/or dates were included in the model, they and all associated two- and three-way interactions were analyzed as within-plot repeated measures and were tested using full model residuals. Mean separation procedures were largely least significant ($P < 0.05$) difference contrasts.

Water yield and sediment production data were summarized using arithmetic means and standard deviations only. Data were not subjected to more sophisticated analyses because of the skewed effects of zeros in data sets (i.e., limited precipitation and/or runoff).

Results and Discussion

Precipitation and Soil Water

Amounts and patterns of annual precipitation varied widely among years (fig. 7.1). Annual precipitation during the pretreatment year of 1993 totaled 47.1 cm, 34 percent above the norm of 34.1 cm. Annual precipitation in 1994 was 24.7 cm, 24 percent below normal with only 13.5 cm received during the period from 1 June to 31 October. Total precipitation received on the drought plots was 10.7 cm in 1994 with <0.5 cm received during the period from 1 June to 15 October.

Soil water content was greatest during 1993, the highest rainfall year, and extended to deeper depths (fig. 7.2). Surprisingly, soil water content did not differ between the drought and nondrought treatment lysimeters. This was because the impact of the ongoing regional drought was near equal to that of the imposed drought. Likewise, grazing treatments did not alter soil water content. An unexpected result was the gradual increase in soil water over years at the 120 cm depth (fig. 7.2). We suspect this was because of downward leakage of water around the soil water monitoring access tubes, as they passed through the near impervious clay pan about 1 m below the soil surface.

Herbage Dynamics and Net Primary Productivity

As expected, harvested biomass estimates varied among years and dates (fig. 7.3) because of temporal variation in annual growing conditions, particularly precipitation (fig. 7.1). In addition, analyses of pretreatment biomass estimates (i.e., 1993) showed there were significantly lesser amounts of cool-season perennial grasses in the drought than nondrought plots (572 vs. 1,237 kg ha^{-1}) resulting in lesser amounts of total biomass (2,018 vs. 2,331 kg ha^{-1}).

In 1994, harvested biomass averaged 349 kg ha^{-1} less in the drought than the nondrought plots (1,735 vs. 2,084 kg ha^{-1}), and it averaged about 1,000 kg ha^{-1} less in the two grazed treatments than in the ungrazed treatment (fig. 7.3). This latter effect was because grazing reduced available biomass an average of 1,387 and 778 kg ha^{-1}, respectively, during the early June and July grazing events. The year following the imposed drought (i.e., 1995), average biomass was less in the drought than nondrought plots (1,259 vs. 1,673 kg ha^{-1}) and varied significantly among all three grazing treatments, averaging 782, 1,422, and 2,196 kg ha^{-1} in the 94–95 grazed, 94 grazed, and ungrazed treatments, respectively. The drought plot effects were largely the result of pretreatment (i.e., 1993) differences in herbage biomass between 1994 drought (2,018 kg ha^{-1}) and nondrought (2,331 kg ha^{-1}) lysimeters. The differences among grazing treatments were a reflection of the effects of both the 1994 and 1995 grazing events. In 1996, the first postdrought year when no plots were grazed, only the main effects of grazing and date were significant. The grazing effect was a carry-over effect of the 1995 grazing events, in that average biomass in the 94–95 grazed treatment was less than in the 94 grazed and ungrazed treatments (1,398 vs. 2,353 kg ha^{-1}). The date effect was a reflection, as in previous years, of normal seasonal growth dynamics and was typical for Northern Great Plains grasslands (Sims and Singh 1978; Singh et al. 1982; Heitschmidt et al. 1995).

Pretreatment aboveground net primary productivity (ANPP) estimates showed productivity was similar across plots (average, 3,096 kg ha^{-1}), although

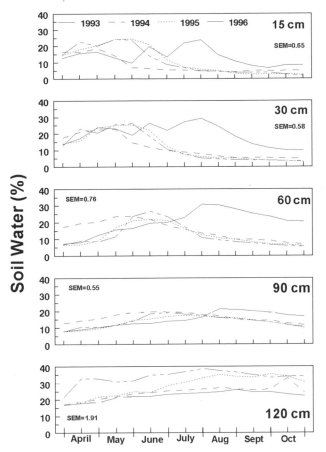

Figure 7.2. Percentage of soil water during the 1993 through 1996 growing season at five depths averaged across drought and grazing treatments. (from Heitschmidt et al. 1999)

the amount of cool-season perennial grass was 657 kg ha[-1] less in the drought than nondrought plots (table 7.1). From 1994 through 1996, average ANPP across plots was 2,282 kg ha[-1], was greatest in 1994 (2,650 kg ha[-1]) and least in 1996 (1,975 kg ha[-1]), was greater in the nondrought than drought plots (2,048 vs. 2,517 kg ha[-1]), and was not affected by grazing treatment. Although minor changes in plant species composition occurred during the four-year study, composition was essentially the same in 1996 as in 1993.

The results of the ANPP component of this study support the general conclusion that grazing is a secondary factor affecting Northern Great Plains ecosystem processes, whereas drought is a primary factor (Reed and Peterson 1961; Olson et al. 1985; Biondini and Manske 1996; Biondini et al. 1998). Even

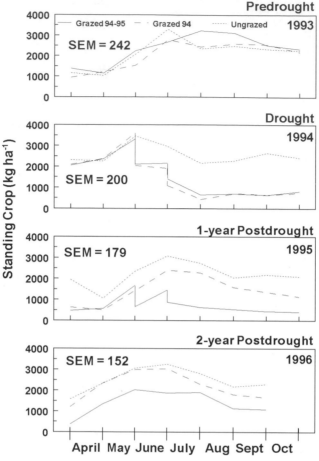

Figure 7.3. Herbage standing crops (kg ha⁻¹) for the three grazing treatments for 1993 through 1996 averaged across drought treatments. (from Heitschmidt et al. 1999)

so, the documented impact of the imposed drought in this study was certainly less than expected, although in retrospect not unexplainable considering the seasonality of the drought. These grasslands are dominated by cool-season plant species that complete most of their growth by late spring and early summer (Sims and Singh 1978; Singh et al. 1982; Dodd et al. 1982; Heitschmidt et al. 1995); thus, plants only need sufficient amounts of soil water until late spring to complete their normal production cycle. By initiating the drought in late spring, most of the annual production cycle was either already completed and/or sufficient soil water was present in the drought plots (fig. 7.2) to permit completion of the cycle with minimal impact. An additional

TABLE 7.1.
Estimated aboveground net primary productivity (kg ha^{-1}) by drought treatment (D = drought, N = nondrought), year, and functional group (cspg = cool-season perennial grass, wspg = warm-season perennial grass, csag = cool-season annual grass).

Species Group	1993		1994		1995		1996	
	D	N	D	N	D	N	D	N
cspg[1]	1030$_a$	1687$_b$	921$_a$	1896$_b$	1106$_a$	1623$_b$	999$_a$	1367$_b$
wspg	1457	1377	1287$_a$	933$_b$	707	584	575	397
csag	199	139	18	16	27$_a$	219$_b$	138$_a$	256$_b$
Forbs	217	86	123	106	99	79	144	74
Total	2903	3289	2349$_a$	2951$_b$	1939$_a$	2505$_b$	1856	2094

[1] Means within a row and within a year, means followed by different letters are significantly different at $P < 0.05$.

factor suppressing the magnitude of drought effect during 1994 was that the nondrought plots were also subjected to drought conditions (fig. 7.1), although severity was less than that imposed on the drought treatment plots. As a result of these two conditions, detection of the imposed drought effects on plant species composition and postdrought recovery patterns was most likely impaired. Further evidence in support of this conclusion can be garnered from other regional studies, wherein annual productivity estimates were similar to those reported in this chapter (e.g., see Singh et al. 1982; Heitschmidt et al. 1995; Biondini and Manske 1996).

Tiller Growth Dynamics

Severity of defoliation (i.e., percentage utilization by weight) of blue grama tillers per grazing event averaged 40 percent with few differences between drought treatments, among grazing treatments, or between years. In 1994, the year of the imposed drought, rgrs were greater during spring than summer (table 7.2) regardless of drought treatment. These differences reflect normal growth patterns for this region, as growth is most rapid during spring followed by reduced growth during summer as plants mature. The season by grazing treatment interaction effect was the result of accelerated rgr occurring in both grazed treatments following the June grazing event (i.e., late spring). Tiller agrs (data not shown) were affected by drought treatment, being 38 percent greater in the nondrought than drought treatment, unaffected by grazing treatment, and affected by season. Seasonal effects were again related to expected changes in growth rates over a growing season.

There were no interaction effects. Relative to other variables examined in 1994, the imposed drought reduced emergence of blue grama axillary tillers by 32 percent (8 percent vs. 40 percent) when averaged across grazing treatments, whereas emergence in the grazed treatments was 22 percent greater (32 percent vs. 10 percent) than in the ungrazed when averaged across the drought treatments. However, the accelerated tiller RGR and emergence of axillary tillers associated with the grazed treatments, and the suppressed AGR and emergence of axillary tillers associated with the drought treatments, did not translate into differences in estimated ANPP for blue grama. No pretreatment differences (i.e., May 1994) in tiller densities were found between lysimeters, regardless of assigned drought or grazing treatment.

In 1995, the year following the imposed drought, RGRs were affected by drought treatment, grazing treatment, and season (table 7.2). However, unambiguous data interpretation was difficult because all two-way interactions were also significant. But analyses showed generally that (1) RGRs in the drought treatments were slightly greater than in the nondrought treatments; (2) grazing stimulated growth during late spring; and (3) seasonal effects were linked to expected seasonal changes in plant growth rates. In addition, results indicated that neither growing season rates of axillary tiller emergence, nor May 1994 to May 1995 tiller replacement rates, nor ANPP were affected by the previous year's drought treatment. Grazing again enhanced axillary tiller emergence rates, averaging 60 percent in the 94–95 grazed treatment as compared with 47 percent and 30 percent in the 94 grazed and the ungrazed treatments, respectively. No differences in tiller densities were found among treatments in May 1995.

Severity of defoliation of western wheatgrass tillers was generally greater than for blue grama tillers, averaging 50 percent as compared with 40 percent for blue grama. In general, tiller response patterns were similar to blue grama with some notable exceptions. In 1994, both tiller RGR (table 7.2) and AGR were enhanced following the June grazing event. Emergence of axillary tillers was reduced from 79 percent to 7 percent by the drought treatment and enhanced 11 percent by grazing (46 percent vs. 35 percent). No differences were found between drought or among grazing treatments relative to pretreatment tiller densities. However, western wheatgrass ANPP was significantly greater in the nondrought than drought treatments (1,450 vs. 800 kg ha^{-1}).

In 1995, RGR (table 7.2) and AGR were again unaffected by the 1994 drought treatment, but they were affected by grazing treatment and season

TABLE 7.2.
Mean relative growth rates (RGR = gm gm^{-1} da^{-1}) for blue grama and western wheatgrass tillers in 1994 and 1995 by drought treatment, grazing treatment, and season.

Variable	DROUGHT TREATMENT[1]		GRAZING TREATMENT			SEASON		
	Drought	Nondrought	94–95 Graze	94 Graze	Ungrazed	Mid-spring	Late spring	Summer
Blue grama								
1994[2]	0.006	0.011	0.011	0.012	0.003	0.014$_a$	0.012$_a$	-0.001$_b$
1995[2,3,4]	0.021$_a$	0.018$_b$	0.022$_a$	0.014$_b$	0.021$_a$	0.039$_a$	0.018$_b$	0.000$_c$
Western wheatgrass								
1994[2]	0.006	0.005	0.008$_a$	0.006$_a$	0.002$_b$	0.008$_a$	0.008$_a$	-0.005$_b$
1995[2]	0.011	0.011	0.017$_a$	0.008$_b$	0.007$_b$	0.016$_a$	0.016$_a$	0.002$_b$

[1] Within a row and within a main effect, means followed by different letters are significantly different at $P < 0.05$.
[2] Significant grazing treatment by season interaction.
[3] Significant grazing treatment by drought interaction.
[4] Significant drought by season interaction.

and their interaction. The interaction was because RGRS were greater in the 94-95 grazed treatment than the 94 grazed and ungrazed treatments in mid-spring and greater in the 94-95 and 94 grazed treatments in late spring than the ungrazed treatment. There were no differences among treatments during summer. Relative to other measured variables, growing season axillary tiller emergence rates and May 1994 to May 1995 tiller replacement rates were unaffected by either drought or grazing treatments, and ANPP was again greater in the nondrought than drought treatment plots (1,650 vs. 1,175 kg ha⁻¹).

The detailed results of these studies are complex and difficult to interpret; yet general responses follow those reported by others. For example, it is well known that drought reduces plant growth rates as reflected by curtailed rates of leaf extension (e.g., see Busso and Richards 1993; Hsiao et al. 1985), and reductions in number of leaves (Reed and Peterson 1961), size of lamina (Nesmith and Richie 1992), and number of flowering stalks (Coupland 1958) and tillers (Haferkamp et al. 1992; Zhang and Romo 1995). However, the effects of grazing are not as clear as drought; some studies have shown defoliation enhances tiller growth of blue grama (Reece et al. 1988) and northern wheatgrass (Zhang and Romo 1995) when growing conditions are favorable (e.g., early June in this study), while others have shown grazing/defoliation often inhibits tiller growth (Stout et al. 1980; Olson and Richards 1988; Butler and Briske 1988; Murphy and Briske 1992, 1994). But regardless of the plant growth processes impacted by the drought and grazing treatments, effects were apparently of insufficient magnitude to significantly alter community level response patterns relative to primary productivity and plant species composition (Heitschmidt et al. 1999).

Root Growth Dynamics

Analyses of total number of root intersects (i.e., summed across all depths) revealed that date was the only significant main effect and that the interactions of date with drought and grazing treatments were the only significant interaction effects. However, no clear date-dependent patterns were apparent relative to these effects other than the clear effects of the drought treatment during 1994 (fig. 7.4). A detailed examination of these data showed that both pattern and magnitude of variation in total intersects was similar to the pattern and magnitude of intersects in Bw soil horizon (i.e., 20–40 cm depth) alone. The reasons for this are unclear, but it may be related to the findings of Voorhees (1989), Allmaras and Logsdon (1990), Huck and Hoogenboom

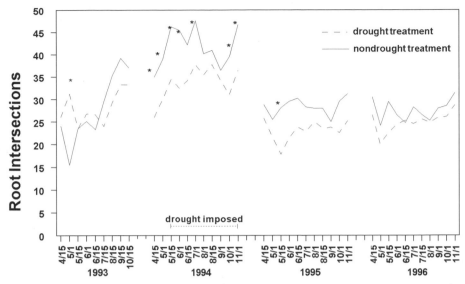

Figure 7.4. Total number of observed roots by drought treatment for 1993–1996 averaged across grazing treatments. Means within a date are significantly different at $P < 0.05$ if asterisk present. (from Hild et al. 2001)

(1990), Zobel (1991), and Jesko et al. (1997), wherein they report that compensatory shifts in root growth occur among sites (e.g., soil horizons) depending upon growing conditions. Although our drought treatment did not significantly alter soil water content in this study (fig. 7.2), it is quite possible that subtle, statistically nonsignificant differences in soil water in the Bw horizon caused significant changes in root growth.

Xylem Water Potential
Drought treatment did not affect water potentials of either blue grama or western wheatgrass plants either year. As expected, water potentials were generally greater at mid-day (-2.33 MPa) than predawn (-1.45 MPa), with little difference between western wheatgrass (-1.98 MPa) and blue grama (-1.80 MPa), regardless of time of day. However, temporal changes between years in predawn water potentials clearly showed the water stress impact of both the imposed drought and the regional drought of 1994. For example, in 1994, predawn water potentials changed from an average of -1.07 MPa on 23 June to -4.11 MPa on 30 June, reaching a maximum of -4.73 MPa on the last sample date of 28 July.

This is in contrast to 1995, when the change in stress from -1.42 MPa to

-4.07 MPa was delayed until between 17 July and 2 August, the last sample date. These data provide further support for the conclusion that the 1994 nondrought plots were experiencing drought stress near equal to that in the imposed drought plots.

Seed Bank

The number of emerging seedlings was dominated by cool-season annual grasses (table 7.3). The total number of cool-season annual grass seedlings was similar between drought treatments in 1993, the year before the imposed drought, and in 1996. However, in 1994, the number of emerged seedlings was greater in the nondrought than drought plots, whereas the opposite was true in 1995. We hypothesize that the 1994 results were related to the effects of the drought on seed viability rather than seed production, because the drought was not initiated until late May, after the annual cool-season grasses had completed their life cycles (Karl et al. 1999). We further hypothesize that the 1995 results reflected the impacts of secondary dormancy processes on the overwhelmingly dominant annual cool-season grass, Japanese brome, as described by Baskin and Baskin (1981) and Haferkamp et al. (1994, 1995, 1996). Based on their findings, it can be reasoned that the extended drought delayed germination until air temperatures were sufficiently low to induce secondary dormancy. Thus, the crop of annual grass seeds was most likely greater in the drought than nondrought plots in 1995, because 1994 dormant seeds were carried over.

The second most abundant group of emerging seedlings were forbs (table 7.3). Averaged across grazing treatments and years, the number of forb seedlings was greater in the nondrought than drought plots (206 vs. 126 seedlings m^{-2}). However, the number of emerging forbs was less in the drought grazed (123 seedlings m^{-2}) than the nondrought grazed (251 seedlings m^{-2}) plots, with no differences between the drought and nondrought ungrazed treatment plots (122 seedlings m^{-2}). Because species composition of the emerging population of forb seedlings did not vary significantly among years or treatments, it is difficult to clearly identify causal factors associated with forb dynamics. Still, results indicate that grazing may increase the number of emerging forb seedlings when sufficient water is available, whereas declines can be expected with drought regardless of grazing treatment.

Cool- and warm-season perennial grasses were both marginally ($P = 0.13$ and $P = 0.08$) affected by drought, unaffected by grazing treatment, and significantly affected by year (table 7.3). However, when summed across

TABLE 7.3.
Numbers of emerging seedlings (m^{-2}) for drought treatments (D = drought, N = nondrought) during 1993–1996 averaged across grazing treatments.

Species Group	1993 D	1993 N	1994 D	1994 N	1995 D	1995 N	1996 D	1996 N
Perennial grasses	2.4	12.8	16.3a[1]	44.7b	14.2	21.6	36.1a	77.1b
Cool-season	1.6	2.5	3.8	10.5	9.5	11.3	8.3	33.8
Warm-season	0.8	10.3	12.5	34.2	4.7	10.3	27.8	43.3
Cool-season annual grasses	318.3	313.3	240.3a	535.5b	596.7a	238.3b	239.7	256.3
Forbs	32.5	38.6	126.2a	201.3b	185.8a	222.8a	160.8a	360.0b
Total	353.2	364.7	382.8a	781.5b	796.7a	482.7b	436.6a	693.4b

[1] Within a row and within a year, means followed by different letters are significantly different at $P < 0.05$.

functional group, total emerging perennial grass seedlings were significantly affected by drought, year, and the interactions of drought and year and grazing treatment and year. Averaged across grazing treatments and years, perennial grass seedlings averaged 17 and 39 m^{-2} in the drought and nondrought plots, respectively, but were only significantly different in 1994 and 1996. The grazing by year interaction effect was caused by a greater number of seedlings in 1996 in the 94–95 grazed than the 94 grazed and ungrazed seedlings treatments (47, 30, and 25 seedlings m^{-2}, respectively). The sparsity of perennial grass seedlings follows the findings of Rice (1989), Simpson et al. (1989), and Kinucan and Smeins (1992).

Water Budget, Sediment Production, and Nutrient Flux

Growing season (i.e., May–October) water budgets were proportionally unaffected by either drought or grazing treatments, with 99 percent of growing season precipitation lost via evapotranspiration regardless of treatment. This estimate of water loss via evapotranspiration processes is greater than normally recorded for semiarid and arid ecosystems (Opperman et al. 1977; Weltz 1987; Carlson et al. 1990), but not unexpected considering that (1) this budget is only for a growing season and not an entire year (i.e., no snowmelt over frozen soil); (2) the number of growing season runoff events at this location are few; and (3) deep percolation is not a factor. Drought did quantitatively alter budgets simply through reducing the amount of precipitation. For example, runoff during the imposed drought of 1994 was significantly greater on the nondrought than on the drought plots (0.35 vs. 0.00 cm),

TABLE 7.4.

Sediment and nutrient losses (g ha^{-1}) for drought and nondrought treatment plots in 1994 and 1995 averaged across grazing treatments (from Emmerich and Heitschmidt 2001).

Variable	1994[1]		1995[1]	
	Drought	Nondrought	Drought	Nondrought
Ca	0	10.0	27.0	11.8
Mg	0	4.1	4.8	3.2
K	0	6.4	11.4	7.5
Na	0	2.4	4.7	5.8
NO$_3$-N	0	3.3	1.1	0.9
NH$_3$	0	1.8	0.9	0.6
PO$_4$-P	0	0.2	0.5	0.2
Cl	0	4.1	5.1	3.0
SO$_4$	0	2.2	11.0	4.5
HCO$_3$	0	20.9	46.1	30.1
Sed-N[2]	0	2.8	14.9	8.9
Sed-P[2]	0	0.5	2.9	1.5
Sediment[2]	0	664	5889	2548

[1] All 1994 and no 1995 drought and nondrought values were significantly different at $P < 0.05$.
[2] Significant ($P < 0.05$) grazing by drought treatment interaction effects.

although the amount of runoff per growing season averaged across all years and treatments was only 0.23 cm. Annual amounts of soil water recharge from November 1 to April 30 averaged 1.87 cm and were unaffected by drought or grazing treatments. Similarly, growing season soil water drawdown and eventual loss through evapotranspiration were unaffected by year and treatments, with an average loss of 8.56 cm. Moreover, evapotranspiration was affected by drought treatment and year, averaging 22.9 and 29.5 cm for the drought and nondrought plots, respectively, and 21.0, 27.9, and 29.6 cm during 1994, 1995, and 1996, respectively.

Because runoff was near zero, growing season sediment production was minimal, averaging 1.7 kg ha^{-1} yr^{-1}. Similarly, annual nutrient losses, as monitored during the 1994 and 1995 growing seasons, were minimal, even though nutrient losses on the nondrought plots were significantly greater than on the drought plots during the imposed drought year of 1994 (table 7.4). This was because no runoff occurred on the drought plots as compared to 0.035 cm for the nondrought plots. Although nutrient losses during 1995 tended ($P < 0.20$) to be greater on the drought than on the nondrought plots, the only significant treatment effects were related to grazing. In general, treatment

Table 7.5.
Sediment and nutrient losses (g ha^{-1}) for three grazing treatment plots during 1994 averaged across drought treatments (from Emmerich and Heitschmidt 2002).

Variable	Ungrazed	1994 Grazed	1994–95 Grazed
Ca	0.5$_a$[1]	25.2$_b$	32.5$_b$
Mg	0.1$_a$	4.8$_b$	7.1$_b$
K	0.2$_a$	10.1$_b$	17.8$_b$
Na	0.2$_a$	5.9$_b$	9.6$_b$
NO$_3$-N	0.01$_a$	0.96$_b$	2.08$_b$
NH$_3$	0.01$_a$	0.47$_b$	1.75$_c$
PO$_4$-P	0.001$_a$	0.31$_b$	0.73$_c$
Cl	0.1$_a$	4.5$_b$	7.6$_b$
SO$_4$	0.2$_a$	8.5$_b$	14.6$_b$
HCO$_3$	1.3$_a$	46.9$_b$	66.1$_b$
Sed-N[2]	0.2	12.9	22.6
Sed-P[2]	0.03	2.19	4.35
Sediment[2]	31	3972	8652

[1] Within a row, means followed by different letters are significantly different at $P < 0.05$.

[2] Significant ($P < 0.05$) grazing by drought treatment interaction effects.

plots grazed during both 1994 and 1995 experienced greater nutrient losses than those grazed in 1994 only, which experienced greater losses than the ungrazed plots (table 7.5). These differences were related to statistically insignificant ($P > 0.10$) increasing amounts of runoff associated with statistically significant ($P < 0.05$) decreasing amounts of standing herbage (fig. 7.3). The impacts of grazing on water yield, sediment production, and ultimately nutrient transport were probably dampened in this study because of the limited amount of ambient precipitation received in 1994. Others have reported that grazing dramatically increases runoff and sediment production in semiarid grassland ecosystems (Dadkhah and Gifford 1980; Thurow et al. 1988; Thurow 1991). However, even with accelerated rates of runoff and sediment production, nutrient transport from these grasslands is low (Neff 1982) and similar to other arid and semiarid grasslands (Owens et al. 1983; Barros et al. 1995).

Conclusions

Although the results of this study are confounded by the presence of a natural drought during the summer of 1994, we conclude that increased intensity of summer drought will have only limited impact on these grasslands regardless of grazing regimen. Summer drought is normal for this region of the Northern Great Plains (fig. 7.1), and its impact is dampened considerably by the amount of stored soil water (fig. 7.2) from fall and spring precipitation events (fig. 7.1). We would hypothesize, therefore, that a shift in drought conditions toward spring would most likely impact the structure and function of these grasslands considerably more than the severe summer/fall drought conditions examined in this study. Research is currently underway at this location to test the spring drought hypothesis.

Acknowledgments

Research was conducted under a cooperative agreement between USDA–ARS and the Montana Agricultural Experiment Station. The mention of a proprietary product does not constitute a guarantee or warranty of the product by USDA, Montana Agricultural Experiment Station, or the author and does not imply its approval to the exclusion of other products that may also be suitable. USDA–Agricultural Research Service, Northern Plains Area, is an equal opportunity/affirmative action employer and all agency services are available without discrimination.

8

Response of Southwestern Oak Savannas to Potential Future Precipitation Regimes

JAKE F. WELTZIN & GUY R. MCPHERSON

Predicted changes in global and regional precipitation regimes (as described in chapter 1) are expected to have important ramifications for the composition and diversity of plant communities and ecosystems (e.g., VEMAP Members 1995; Neilson and Drapek 1998; Cramer et al. 2001). In arid and semiarid regions where vegetation is highly dependent upon precipitation, changes in seasonal precipitation and soil moisture regimes may cause major shifts in plant composition, distribution, and abundance (Neilson 1986; Stephenson 1990).

One mechanism by which arid-region plant communities might respond to shifts in the amount and seasonality of precipitation is the process of soil moisture resource partitioning. According to this model, or "two-layer hypothesis" (Walter 1954, 1979), grasses use shallow sources of soil moisture derived from summer precipitation, whereas deep-rooted woody plants use precipitation that percolates deep into the soil profile during the nongrowing season. This process of niche differentiation, or soil moisture partitioning, has been widely invoked to explain coexistence of grasses and trees in savannas and woodlands throughout the world (Knoop and Walker 1985; Sala et al. 1989; Brown and Archer 1990; Peláez et al. 1994; Schulze et al. 1996).

If soil moisture partitioning constrains the structure of plant communities, then potential increases in summer precipitation (Neilson and Drapek 1998) should favor shallow-rooted species and "warm-season" grasses with the C_4 photosynthetic pathway (e.g., Knoop and Walker 1985; Ehleringer et al. 1991; Lauenroth et al. 1993; Burgess 1995). Alternatively, potential increases in winter soil moisture content (Manabe and Wetherald 1986) may favor "cool-season" herbaceous plants and woody plants, most of which have the C_3 photosynthetic pathway. Concordant with this hypothesis, increases in winter precipitation have been observed to facilitate growth and establishment of woody plants within semiarid grasslands (Brown et al. 1997).

In this chapter, we examine how changes in the amount and seasonality of precipitation may affect the structure and function of oak savannas characteristic of the southwestern United States and northwestern Mexico. These oak savannas are located at the ecotone between temperate oak woodland and adjacent semidesert grassland, although the character of this ecotone can range from discrete lower tree lines to relatively diffuse savannas (Brown 1982; McClaran and McPherson 1999). Precipitation regimes in this region are bimodal, with peaks in quantity of precipitation during summer and winter. However, this regional precipitation regime is likely to change within the next century as atmospheric CO_2 concentration increases (Giorgi et al. 1998; Houghton et al. 2001). We used manipulative experiments to simulate potential scenarios of precipitation redistribution that this plant community may experience by the mid- to late twenty-first century, and we focused on the response of the herbaceous plant community and the recruitment and production of oak (*Quercus* L.) seedlings.

Methods

Study Site

Research was conducted between 1994 and 1996 at the lower (and drier) margin of temperate, evergreen oak woodland at the base of the Huachuca Mountains in southeastern Arizona. Savannas dominated by *Quercus emoryi* Torr. (Emory oak) characterize the ecotone between oak woodland and adjacent semidesert grassland, which is dominated by C_4 perennial bunchgrasses (Brown 1982; McClaran and McPherson 1999).

The study site was located in lower Garden Canyon (31°29' N, 110°20' W) on Fort Huachuca Military Reservation near Sierra Vista, Arizona. Overstory tree cover within the savanna was 11 percent, as estimated from aerial photographs (Haworth and McPherson 1994). Herbaceous vegetation was dominated by the perennial C_4 bunchgrass *Trachypogon montufari* (H.B.K.) Nees (crinkleawn). The site is 1,550 m in elevation with a 5 percent slope on a northeastern aspect. Soils developed from gravelly alluvium. Climate is semiarid, with an average annual temperature of 20°C. Average annual precipitation is 602 mm, and is bimodally distributed, with peaks during the summer monsoon (July–September; 50 percent) and during winter (December–March; 30 percent) (NOAA 1996). Weltzin and McPherson (1999, 2000) provide further details on climate, vegetation, and soils at this site.

Selection of Treatments Vis-à-Vis Potential Future Precipitation Regimes
Most general circulation models are unable to make concrete predictions as to how precipitation regimes will change in the near future on regional scales (Mahlman 1997). First, their scale of resolution may be too coarse: a typical general circulation model grid may encompass an area the size of Arizona. Second, multiple factors may interact and affect precipitation at several scales, which adds to the complexity of models. For example, in the southwestern United States, precipitation regimes may be affected by changes in the intensity of convective activity associated with the Bermuda High (with attendant effects on summer precipitation) and changes in mid-latitude jet streams, frontal systems, and/or the presence and strength of the El Niño Southern Oscillation (which may alter winter precipitation) (Leverson 1997; Giorgi et al. 1998). Finally, the topographic complexity of this region further complicates predictions of potential future precipitation regimes, particularly on local scales.

Given these limitations, a nested regional climate model developed by the National Center for Atmospheric Research has predicted that a doubling of current CO_2 concentration will decrease the amount of winter precipitation in the southwestern United States (Giorgi et al. 1998). Further, increases in summer precipitation that increase the density and production of C_4 grasses have been predicted to contribute to substantial "greening" of the region (Neilson and Drapek 1998; Neilson, chapter 4, this volume). However, in contrast to these predictions, the HADCM2 general circulation model developed by the Hadley Centre at the U.K. Meteorological Office predicts that the Southwest will experience both drier summers and wetter winters by the year 2030. In sum, precipitation regimes in this region are expected to remain bimodal, but models predict either increased or decreased precipitation in the summer or winter.

Therefore, one goal of our experiment was to test the sensitivity and range of the response of southwestern oak savannas to several different scenarios of climate change. We focused on changes in both summer and winter precipitation because (1) the majority of precipitation is currently received in these two seasons, (2) models predict that summer and/or winter precipitation are likely to change (albeit in uncertain amounts and directions), and (3) changes in the relative proportion of seasonal precipitation may have important ramifications for the structure and function of ecosystems (see also, in this volume, Neilson, chapter 4; Svejcar et al., chapter 6; Heitschmidt and Haferkamp, chapter 7).

Experimental Design

In June 1994, we initiated a field experiment consisting of five simulated precipitation treatments applied to plots isolated from ambient precipitation and soil moisture (fig. 8.1). The first treatment received simulated precipitation equivalent to the long-term (i.e., 30-year) mean annual precipitation for the site (602 mm/yr) (table 8.1). The other four treatments received all possible combinations of 50 percent additions and reductions of summer (July–September) and winter (December–February) precipitation relative to the long-term seasonal mean. Treatments received equal amounts of precipitation in spring (March–June) and autumn (October–November). This experimental design necessarily confounded seasonal treatments with total precipitation; equivalent proportional reductions or additions to both summer and winter mean precipitation, which currently differ by a factor of almost two, result in a broad range of total annual precipitation (table 8.1).

Treatments were arranged within a randomized complete block design (n = 4). Blocks were established within homogeneous stands of perennial bunchgrasses. Within each block, five 1.2 m by 1.5 m plots were arranged linearly at 1.5 m spacing. The perimeter of each plot was trenched to 1 m depth and lined with polyethylene film to prevent lateral movement of soil water. The edge of each plot was bordered to prevent lateral movement of surface water. Vegetation in each plot was left intact. A permanent precipitation shelter (16 m by 4 m) constructed of steel tubing, clear polyethylene film, and fence posts was erected over each block to exclude ambient precipitation (fig. 8.1). The pitched roof of each shelter was 2.2 m above ground level at its apex and 1.5 m high along the sides and ends. Poultry netting (2.5 cm mesh) was wired to fence posts and rebar stakes around each block to form a 60 cm tall vertebrate exclosure.

Shelters were open sided to minimize microclimatic impact. Shelters reduced photosynthetically active photon flux density by 29 percent $\pm$ 10 percent (mean $\pm$ 1 SE) at solar noon on a clear, midsummer day. Although shelters likely altered other, unquantified microenvironmental variables (e.g.,

Figure 8.1. Experimental system for capturing and redistributing precipitation at the south-western oak savanna study site in southeastern Arizona. Precipitation shelters (16 m by 4 m; n = 4) were covered with clear polyethylene film to exclude ambient precipitation from experimental plots. Gutters routed ambient precipitation to storage tanks. Collected precipitation was applied to plots according to a protocol designed to simulate natural precipitation patterns.

ambient temperature, relative humidity), experimental units were affected equally (see discussion of shelter design and construction in chapter 4, this volume).

Precipitation collected and stored on-site was applied to plots according to a randomly generated precipitation regime that simulated natural precipitation patterns (CLIGEN, USDA–ARS Southwestern Watershed Research Center, Nicks and Lane 1989). Simulated precipitation events ranged from 1 mm to 120 mm, and were applied by hand-watering 57 times annually (table 8.1).

Gravimetric soil moisture content at 10 cm and 50 cm in each plot was determined monthly by extracting a ~20 g soil sample using a 2 cm diameter coring tool. Core holes were backfilled with tamped soil. Results of soil moisture monitoring are provided by Weltzin and McPherson (2000). Generally, statistical contrasts (i.e., summer-wet vs. summer-dry, and winter-wet vs. winter-dry treatments) confirmed that soil moisture reflected the different watering treatments established and applied within the context of the experimental design.

Insecticide (Carbaryl 4L) was applied to all seedlings monthly during the

TABLE 8.1.
Season, frequency of application, and amount (mm) of five precipitation treatments (n = 4) applied to plots isolated from ambient precipitation and soil moisture at a southwestern oak savanna site in southeastern Arizona. Long-term mean represents the 30-year average seasonal precipitation for the site, and seasonal wet and dry treatments represent 50 percent additions and reductions, respectively, of the long-term seasonal mean.

Season	Months	Frequency	Long-term Mean	TREATMENT			
				Summer-Dry Winter-Wet	Summer-Dry Winter-Dry	Summer-Wet Winter-Wet	Summer-Wet Winter-Dry
Spring	March, April, May, June	7	62	62	62	62	62
Summer	July, August, September	29	315	158	158	473	473
Autumn	October, November	7	53	53	53	53	53
Winter	December, January, February	14	172	258	86	258	86
Total		57	602	531	359	846	674

growing season to minimize invertebrate herbivory. Additional details of the experimental design are provided by Weltzin and McPherson (2000).

Q. Emoryi *Seedling Demography and Production*

On 19 July 1994 and 17 July 1995, coincident with the onset of the summer rains (i.e., the Arizona "monsoon"), we planted 49 *Q. emoryi* acorns at 10 cm spacing into each plot. Acorns were collected on the day of planting from at least 20 trees on-site and were sorted by visual examination and flotation (Nyandiga and McPherson 1992).

For each cohort, seedling emergence and survival were monitored at two-week intervals for the first four months. Thereafter, seedling survival was monitored monthly until experiment termination in October 1996.

At experiment termination, we recorded the height of live *Q. emoryi* seedlings in each cohort, and clipped them off at ground level for determination of aboveground biomass. We then used shovels to excavate each plot to 1 m depth to determine *Q. emoryi* root biomass (in 20 cm increments) and taproot length for two seedlings selected at random (and treated as subsamples) from each cohort. Seedling roots were extracted by hand. All plant tissue samples were oven-dried at 60°C to constant mass before weighing.

Seedling emergence, survivorship, and recruitment were analyzed using analysis of variance (ANOVA) models, proportional hazards regression analysis, and Z-tests of proportions, respectively. Seedling size and biomass data were analyzed using ANOVA. Additional details on statistical analyses are provided by Weltzin and McPherson (2000).

Herbaceous Plant Community Structure and Production

At experiment termination, we determined aboveground herbaceous biomass by clipping each species (to 2 cm height) within a square 0.25 m² subplot centered within each plot. Belowground herbaceous biomass was determined by extracting soil with a bucket auger from two randomly located 10 cm diameter cores in successive 20 cm increments to 1 m depth. Herbaceous roots were extracted by hand separation, flotation in brine, and subsequent sieving through 2.0 mm mesh.

Species richness (S), Shannon's diversity index (H'; Shannon and Weaver 1949), and the reciprocal of Simpson's (1949) index (1/D) were used to characterize herbaceous communities. H' and 1/D are nonparametric indices (Southwood 1978) that incorporate both richness and evenness (Peet 1974). We used ANOVA models to evaluate block and treatment effects on biomass of species and functional groups, and S, H', and 1/D. In addition, we used single degree of freedom contrasts to test for differences between specific treatment combinations (e.g., summer-wet vs. summer-dry; Zar 1996). *P* values are for ANOVAS unless otherwise indicated. We also used Pearson product-moment correlation coefficients to assess correlation between total annual precipitation and C_3, C_4, and total production.

Q. Emoryi *and* T. Montufari *Water Potential*

We determined predawn leaf water potential (Ψ) for one *T. montufari* plant and one *Q. emoryi* plant selected at random within each plot on 20 April, 30 June, 22 August, and 17 October 1996. We determined Ψ with a Scholander-type pressure chamber (PMS Instrument Company, Corvallis, Oregon), using representative leaves collected one to three hours before the beginning of the daily photoperiod. For each sample date, we used an ANOVA model to evaluate effects of block and treatment on Ψ, and single degree of freedom contrasts to compare specific groups of treatments.

Results

Q. Emoryi *Seedling Demography and Production*

Seedling emergence rates were 1.4 to 3 times greater in summer-wet compared with summer-dry treatments, depending on cohort (table 8.2; Weltzin and McPherson 2000). Subsequent survival of emerged seedlings differed little between treatments for both seedling cohorts (table 8.2; fig. 8.2; Weltzin and McPherson 2000). Generally, seedling survivorship curves were characterized by low mortality during the summer "monsoon" in July–September and between November and March (the nongrowing season), and higher rates of mortality during the severe premonsoon drought in April–June and the less-severe postmonsoon drought in October–November. Mortality rates for both cohorts were relatively high during the particularly severe premonsoon drought of 1996. Nonetheless, seedling survival at experiment termination averaged 52 percent and 56 percent for two-year-old and three-year-old seedlings, respectively.

Seedling recruitment reflected the combined effects of emergence and subsequent survivorship but was more heavily weighted by seedling emergence. Recruitment ranged from 1.5 to 3 times greater in summer-wet than summer-dry treatments for both cohorts. Recruitment for cohort 1 was least in summer-dry treatments, intermediate in the mean treatment, and greatest

TABLE 8.2.
Emergence, survival at experiment termination, and recruitment of *Q. emoryi* seedlings planted in 1994 (Cohort 1) and 1995 (Cohort 2) in plots (n = 4) watered to simulate mean seasonal precipitation (Mean) and all combinations of 50 percent additions (Wet) and reductions (Dry) to summer (July, August, September) and winter (December, January, February) precipitation relative to the seasonal mean.

TREATMENT		COHORT 1			COHORT 2		
Summer	Winter	Emergence (%)	Survival (%)	Recruitment (%)	Emergence (%)	Survival (%)	Recruitment (%)
Mean	Mean	50.8 a[1]	47.7 a	24.2 a	47.1 a	35.0 a	16.5 a
Dry	Wet	19.9 b	54.2 ab	10.8 b	32.4 a	57.8 b	18.7 a
Dry	Dry	21.1 b	54.3 ab	11.5 b	38.7 a	43.7 ab	16.9 a
Wet	Wet	59.7 a	61.1 ab	36.5 c	52.1 a	63.9 b	33.3 b
Wet	Dry	62.2 a	61.8 b	38.4 c	48.2 a	58.8 b	28.3 b

[1] Within a column, means follwed by the same letter did not differ ($P > 0.05$) within cohort for emergence (Pr > F), survival (Pr > Z), and recruitment (Pr > Z).

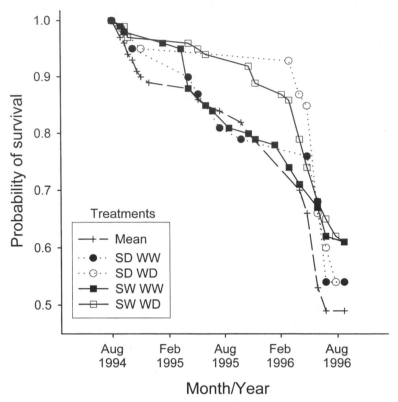

Figure 8.2. Survivorship curves of *Q. emoryi* seedlings planted on 19 July 1994 (i.e., cohort 1). Acorns were planted into experimental plots (n = 4) watered to simulate mean seasonal precipitation (Mean) and combinations of 50 percent additions (W = wet) and reductions (D = dry) to summer (S = July–September) and winter (W = December–February) precipitation relative to the seasonal mean (e.g., SDWW = summer dry, winter wet). Survivorship curves for cohort 2 were similar, and are provided by Weltzin and McPherson (2000). Figure redrawn from Weltzin and McPherson (2000).

in summer-wet treatments ($P < 0.004$). Recruitment for cohort 2 did not differ between the mean and summer-dry treatments ($P > 0.65$), and was greatest in summer-wet treatments ($P < 0.04$).

Seedling size and production parameters at experiment termination (shoot height; root length; and shoot, root, and total biomass) were unrelated to the different treatments (Weltzin and McPherson 2000). Generally, seedling root biomass was concentrated in the top 20 cm of the soil profile (fig. 8.3), and seedling roots extended on average 36 cm to 53 cm deep in the soil for two-year-old and three-year-old seedlings, respectively.

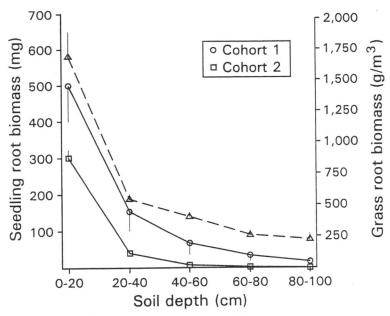

Figure 8.3. Mean root biomass (mg) for *Q. emoryi* seedlings (solid line) planted in July 1994 (cohort 1) and July 1995 (cohort 2), and mean root biomass (g/m³) for *T. montufari* (broken line), in 20 cm soil depth increments at experiment termination in October 1996 (n = 20). Vertical lines represent 1 S.E. Figure from Weltzin and McPherson (2000). Used with permission.

Herbaceous Plant Community Production and Structure

Total herbaceous biomass at the end of the growing season in 1996 was about two times greater in summer-wet than summer-dry treatments (contrast $P <$ 0.0001; table 8.3). Biomass in the mean treatment was greater than in the summer-dry treatments (contrast $P < 0.009$), but did not differ from the summer-wet treatments (contrast $P = 0.09$). Interestingly, total herbaceous biomass was greater in winter-dry than winter-wet treatments (contrast $P = 0.009$).

Biomass of *Trachypogon montufari,* a native C_4 bunchgrass, comprised 66 percent to 88 percent of the total herbaceous biomass, and 87 percent to 99 percent of the C_4 perennial grasses, in all treatments. *T. montufari* biomass was less in summer-dry than summer-wet treatments (contrast $P < 0.001$) and mean treatment (contrast $P < 0.005$); summer-wet and mean treatments did not differ (contrast $P = 0.99$). *T. montufari* biomass did not differ between winter-dry and winter-wet treatments (contrast $P = 0.15$). Biomass of C_3 annual and perennial grasses, and annual and perennial herbaceous dicots, did not differ between treatments ($P > 0.29$).

TABLE 8.3.

Mean (± SE) aboveground herbaceous biomass (g/m²), species richness (S), and diversity indices (1/D, H′). Differences between treatments are described in the text.

Functional Group/Species	Long-term Mean	TREATMENT Summer-Wet Winter-Wet	Summer-Wet Winter-Dry	Summer-Dry Winter-Wet	Summer-Dry Winter-Dry
Perennial C_4 grasses	539 ± 44	499 ± 64	704 ± 111	287 ± 46	316 ± 99
Trachypogon montufari	527 ± 53	468 ± 91	701 ± 112	251 ± 31	298 ± 101
Perennial C_3 grasses	1 ± 1	0 ± 0	0 ± 0	0 ± 0	1 ± 1
Perennial C_3 herbaceous dicots	48 ± 17	80 ± 25	94 ± 43	67 ± 21	50 ± 20
Annual C_3 herbaceous dicots	9 ± 5	27 ± 20	103 ± 59	11 ± 5	84 ± 37
Total herbaceous biomass	597 ± 54	605 ± 51	900 ± 96	364 ± 30	451 ± 75
Species richness, S	5.8 ± 0.3	5.0 ± 0.4	4.5 ± 0.7	5.8 ± 1.1	4.5 ± 0.3
Simpson's index, 1/D	1.3 ± 0.1	1.8 ± 0.5	1.6 ± 0.2	2.0 ± 0.3	2.2 ± 0.5
Shannon's index, H′	0.5 ± 0.1	0.7 ± 0.2	0.6 ± 0.2	0.9 ± 0.1	0.9 ± 0.2

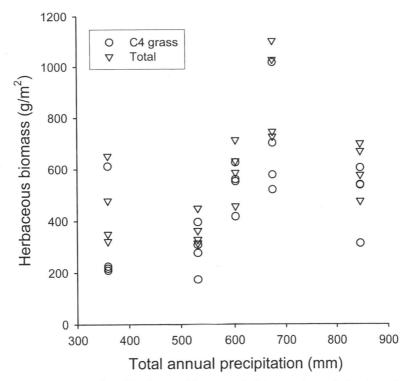

Figure 8.4. C_4 grass and total herbaceous biomass (g/m^2) in experimental plots (n = 4) with different amounts of total annual precipitation (mm) (see table 8.1).

Pearson correlation coefficients indicated that C_3 biomass was independent of total annual precipitation (P = 0.93). Although C_4 (P = 0.04) and total (P = 0.05) biomass covaried with total precipitation, coefficients of determination (r^2) were <0.20 (fig. 8.4).

Total belowground biomass of herbaceous plants at experiment termination was not affected by treatment and averaged 3,100 g/m^3 (Weltzin and McPherson 2000). Herbaceous root biomass was concentrated within the upper 20 cm of the soil profile but extended to depths of at least 1 m (fig. 8.3).

Herbaceous species richness (mean $\pm$ SE = 5.1 $\pm$ 1.3) and evenness (1/D: 1.8 $\pm$ 0.2; H': 0.7 $\pm$ 0.1) did not differ between treatments (P > 0.29).

Q. Emoryi *and* T. Montufari *Water Potential*

Q. emoryi seedling predawn leaf water potentials (Ψ) differed (contrast P = 0.002) between treatments only in October, when Ψ was lower in summer-dry (average, -4.8 MPa) than summer-wet plots (average, -1.9 MPa),

TABLE 8.4.
Predawn leaf water potentials (Ψ; MPa) of *Q. emoryi* seedlings and *T. montufari* in 1996.

Summer	Winter	Q. Emoryi				T. Montufari			
		April	June	August	October	April	June	August	October
Mean	Mean	-1.5[1]	<-6	-0.9	-3.5 a	-1.3	-1.8	-0.2 a	-2.2 a
Dry	Wet	-1.1	<-6	-1.4	-4.2 ab	-0.7	-2.1	-0.3 b	-2.5 a
Dry	Dry	-2.0	<-6	-0.6	-5.4 b	-1.3	-3.5	-0.3 b	-2.3 a
Wet	Wet	-1.4	<-6	-0.6	-1.6 c	-1.0	-2.5	-0.1 a	-0.6 b
Wet	Dry	-2.0	<-6	-0.6	-2.1 c	-1.5	-2.1	-0.2 a	-0.6 b

[1] Means in columns followed by different lowercase letters differed ($P < 0.05$).

and mean treatment plots were intermediate (-3.5 MPa) (table 8.4). Ψ did not differ (contrast $P > 0.05$) between treatments in April (-1.6 MPa) or August (-0.8 MPa). Ψ for all seedlings sampled in June were more negative than -6 MPa, which was the lower limit of measurement for our instrument.

   T. montufari predawn leaf water potentials were less negative in winter-wet (average, -0.9 MPa) than winter-dry (average, -1.4 MPa) treatments in April (contrast $P = 0.01$), did not differ between treatments in June ($P = 0.21$), and were less negative in summer-wet than summer-dry plots in both August and October (contrast $P < 0.002$) (table 8.4). Mean treatment plots did not differ from summer-wet plots in August, and did not differ from summer-dry plots in October (contrast $P > 0.05$).

Discussion

Although changes in regional precipitation regimes are not well predicted by general circulation models, particularly for topographically complex regions such as the southwestern United States, predicted changes in atmospheric circulation and surface temperatures are likely to affect the amount and seasonality of precipitation and soil moisture in this region (e.g., Kattenberg et al. 1996; Giorgi et al. 1998). Results from this study indicate that changes in summer precipitation regimes likely would alter the structure of southwestern oak savannas with respect to both woody plant population dynamics and changes in the herbaceous plant community.

Q. Emoryi *Seedling Demography*
Simulated shifts in the amount and seasonality of precipitation within this oak savanna produced differences in *Q. emoryi* recruitment rates. In particular, summer precipitation was most important for *Q. emoryi* seedling emergence and early establishment, although seedling growth was unaffected. In contrast, recruitment and growth of tree seedlings was uncorrelated with winter precipitation. Ultimately, differences in recruitment among the treatments were shaped primarily by the impacts of summer precipitation on seedling emergence.

   These results are consistent with observations that *Q. emoryi* seedling recruitment is constrained by intra- and interannual variation in summer precipitation patterns (Pase 1969; Neilson and Wullstein 1983; McPherson 1992; Germaine and McPherson 1999; Weltzin and McPherson 2000). Similarly, the distribution and abundance of woody plants in western North

America are often coupled to the quantity and timing of summer precipitation (Neilson and Wullstein 1983, 1985; Ehleringer et al. 1991; Lin et al. 1996; Williams and Ehleringer 2000).

Results from this study indicate that changes in summer precipitation regimes likely would affect *Q. emoryi* population dynamics through changes in seedling recruitment rates. In contrast, shifts in winter precipitation appear to be considerably less important to *Q. emoryi* population dynamics. First, *Q. emoryi* acorns ripen, are dispersed, and emerge coincident with the summer monsoon (McPherson 1992), the importance of which has been demonstrated herein. Second, although seedling, sapling, and adult *Q. emoryi* trees utilize winter precipitation during the early growing season, they are also capable of utilizing summer-derived precipitation even in relatively dry summers (Weltzin and McPherson 1997). However, the importance of winter precipitation to flowering and seed set for this species has not been determined.

Results from this and other studies suggest that once *Q. emoryi* trees become established, they will likely persist in the face of all but extreme environmental change (*sensu* Grime 1979; Warner and Chesson 1985). For example, *Q. emoryi* seedlings were relatively insensitive to environmental conditions imposed in this study (e.g., soil moisture, herbaceous interference), although Germaine and McPherson (1999) found that herbivory and herbaceous competition reduced survival rates of *Q. emoryi* seedlings less than about two years old. Once *Q. emoryi* plants mature into older and larger size classes, they appear to become even more tolerant of environmental perturbation because (1) they are relatively long-lived (Sanchini 1981), (2) they are capable of vegetative reproduction following top removal by fire or drought (Bahre 1991; Caprio and Zwolinski 1995), and (3) they access sources of water from relatively deep in the soil profile (Weltzin and McPherson 1997). This tolerance of unfavorable conditions, coupled with the potential for rapid recruitment during summers with favorable precipitation, suggests that *Q. emoryi* should persist on landscapes where seedling emergence occurs (cf Chesson and Huntley 1989; Archer 1990; McPherson 1997).

Periodic fires historically may have constrained *Quercus* seedling recruitment, although there are few data to support this conjecture (McPherson 1992). However, increases in livestock grazing, habitat fragmentation, and fire suppression activities have reduced the contemporary and future importance of fire as a factor controlling the spatial distribution of this ecotone (Weltzin and McPherson 1995).

Although constraints of seedling recruitment on plant population dynamics are well documented (e.g., Harper 1977; Scholes and Archer 1997), predictions of climate change effects on woody plant abundance and distribution have not always considered this "bottleneck" to woody plant demography (e.g., VEMAP Members 1995; but see Neilson and Wullstein 1983; Neilson 1986; Pastor and Post 1988; Harte and Shaw 1995; Polley et al. 1996). Results of this study suggest that developers of dynamic global vegetation models should incorporate, where possible, empirical data on soil-plant-water relationships that may also include unexpected responses of population processes to changing environmental conditions.

Herbaceous Plant Community Production and Structure

Shifts in seasonal precipitation caused differences in herbaceous community structure and aboveground biomass production after three growing seasons, but did not affect richness or evenness of herbaceous species. Total herbaceous biomass, which consisted largely of C_4 perennial grasses (especially *T. montufari*), increased with increases in summer precipitation. In contrast, total annual precipitation was a relatively poor predictor of C_4 grass and total herbaceous production. Results are consistent with other observations that the production of C_4 grasses, as well as the total herbaceous production of similar systems dominated by C_4 grasses, are correlated with the amount of warm-season precipitation rather than total annual precipitation (Nelson 1934; Cable 1975; Kemp 1983; Burgess 1995).

The individual species or functional group that contributed to greater total biomass in winter-dry than winter-wet treatments could not be determined, because the biomass of each plant species and each functional group did not differ between winter-wet and winter-dry treatments. Within the same region, albeit at a lower elevation, Cable (1975) found that winter precipitation had no consistent effect on perennial grass production the subsequent summer. Because we conducted our sampling at the end of the summer growing season, treatment-induced differences in cool-season plant species production may have been underestimated. Unidentified reductions in production of cool-season species in winter-dry treatments may have removed an interference or competitive barrier (e.g., soil water or nutrient depletion) to subsequent production of warm-season species (Ettershank et al. 1978; Parker et al. 1984; Neilson 1986). Alternatively, dry winters may have altered the composition and abundance of plant pathogens or belowground herbivores (e.g., root-feeding invertebrates) that limit aboveground plant production (e.g., Parker et al. 1984; Bachelet et al. 1989).

Although species diversity in North American grasslands is often related to resource availability (Huston 1994), simulated shifts in seasonal precipitation did not affect richness or evenness of herbaceous species after three growing seasons in this experiment. Dominance of these communities by a single, perennial grass species *(T. montufari)* may have dampened the response of other annual and perennial species to shifts in resource availability *(sensu* Tilman 1993; Silvertown et al. 1994). Similarly, supplemental water increased the abundance of the dominant annual species *(Ambrosia artemesiafolia)* in an old field community, but caused little change in diversity of subdominant species (Goldberg and Miller 1990). In contrast, diversity in communities characterized by codominant annual species (e.g., Mediterranean-type grasslands) with high turnover rates may respond quite rapidly to shifts in abiotic conditions (e.g., Marañón and Bartolome 1993).

Results of this experiment suggest that increases in total herbaceous production may be realized by either decreases in winter precipitation, or especially, increases in summer precipitation. Conversely, reductions in summer soil moisture (e.g., Kattenberg et al. 1996) likely will reduce total herbaceous production through reductions in warm-season grasses (cf Burgess 1995). After three growing seasons, herbaceous community richness and evenness appeared relatively resistant to potential shifts in the seasonality of precipitation.

Soil Moisture Resource Partitioning and Savanna Maintenance

Soil resource partitioning is widely considered a mechanism for stable coexistence of grasses and trees in savannas and woodlands throughout the world (Knoop and Walker 1985; Sala et al. 1989; Brown and Archer 1990; Peláez et al. 1994; Schulze et al. 1996). Similarly, recent research indicates that soil moisture resource partitioning facilitates the coexistence of mature *Q. emoryi* and grasses at this savanna site (Weltzin and McPherson 1997). However, one- and two-year-old *Q. emoryi* seedlings and grasses obtained water from similar depths in the soil profile. This and other results from this study support Weltzin and McPherson's (1997) conclusion that spatial and temporal soil moisture resource partitioning does not occur between *Q. emoryi* seedlings and grasses for at least two years after germination. First, *Q. emoryi* germination, emergence, and early establishment occurred in the summer, when most herbaceous plants were actively growing. This suggests seedling roots can tolerate belowground interference from grass roots long enough to penetrate beyond their influence (Brown and Archer 1990). Second, experimental ma-

nipulations of winter precipitation did not affect recruitment or growth of
Q. emoryi during their first three growing seasons. This suggests that these
seedlings are incapable of using winter-derived moisture from deep in the soil
profile. Third, relatively short root lengths observed for both *Q. emoryi* seedling cohorts, regardless of precipitation treatment, indicate a physical limitation to acquisition of deeper water sources (see also Pase 1969; McPherson
1993).

Results from this study and Weltzin and McPherson (1997) illustrate an
important limitation to the application of the soil resource partitioning model
to explain interactions between woody plants and grasses in this savanna. In
particular, the response of woody plants to environmental conditions is dependent on life-history stages that dictate their belowground morphology.
Consequently, environmental conditions that are sufficient for survival of
adult plants (e.g., adequate winter precipitation) may be insufficient for recruitment of seedlings (cf Harper 1977; Schupp 1995; Weltzin and McPherson
2000), and the response of seedling and mature plants to environmental variation may be decoupled. Accordingly, we propose that the woody plant regeneration niche (Grubb 1977) should be explicitly considered when applying
Walter's (1954, 1979) "two-layer" hypothesis to explain coexistence of woody
plants and grasses.

Spatial and Temporal Dynamics of the Woodland/Grassland Ecotone

Recent research suggests that oak woodland/semidesert grassland ecotones
are stabilized by self-enhancing feedback mechanisms of overstory shade,
seed dispersal, and seedling establishment, coupled with strong abiotic constraints beyond the current ecotone (Germaine and McPherson 1999; Hubbard and McPherson 1999; Weltzin and McPherson 1999). These processes
stabilize the woodland/grassland ecotone both spatially and temporally (cf
Wilson and Agnew 1992). Further, observed (McPherson et al. 1993) and
potential future downslope shifts in the ecotone occur when periodic climatic
conditions simulate, or negate the importance of, conspecific biogenic safe
sites. Results of this research demonstrate that increases in summer precipitation are one mechanism that would facilitate downslope shifts in the wood-
land/grassland ecotone. This interpretation is consistent with that of Mc-
Claran and McPherson (1995), who concluded that the last downslope shift in
this ecotone, which occurred 700–1700 ybp, coincided with a period of particularly high summer precipitation in the region (i.e., the "Medieval Warm"
period, 645–1295 ybp; Davis 1994).

Acknowledgments

We are particularly indebted to Heather Germaine, Andy Hubbard, Mechelle Meixner, Keirith A. Snyder, Jose Villanueva-Diaz, and David Williams. In addition, Laurie Abbott, Debbie Angell, Paulette Ford, Dan Koepke, Anastasia Olander, Sean Schaeffer, and Kim Suedkamp helped characterize the response of the herbaceous plant community. We appreciate the cooperation of Sheridan Stone, Fort Huachuca Office of Game Management. Suggestions from Heather Germaine helped us improve this manuscript.

9

Rainfall Timing, Soil Moisture Dynamics, and Plant Responses in a Mesic Tallgrass Prairie Ecosystem

PHIL A. FAY, ALAN K. KNAPP, JOHN M. BLAIR,

JONATHAN D. CARLISLE, BRETT T. DANNER,

& JAMES K. MCCARRON

Grasslands occur in well-defined climatic zones based on temperature and rainfall regimes (Hayden 1998). The present distribution of North American grasslands is largely the result of climatic zones that developed during the Miocene-Pliocene transition (Borchert 1950; Axelrod 1985). North American grasslands have persisted because of the relative stability of these climatic patterns, and in some cases because of the additional presence of periodic fires (Wells 1970; Knapp, Briggs, et al. 1998)

Evidence is accumulating that climatic patterns in North American grasslands will change in the coming decades, due to continuing increases in the concentration of atmospheric CO_2 and other greenhouse gases (Schneider 1993; Houghton 1997). Predictions from atmospheric general circulation models (GCMS) suggest that grasslands in the Central Plains region will experience higher growing season temperatures and changes in the patterns of growing season rainfall. Predicted changes in rainfall patterns include reduced total growing season rainfall, increased occurrence of convective rainfall events, longer dry intervals between rainfalls, and lower soil moisture (Manabe et al. 1981; Giorgi et al. 1994; Karl et al. 1996; Gregory et al. 1997; Watson et al. 1997).

Most North American ecosystems, including the Central Plains grasslands, are expected to be moderately to highly sensitive to changes in climate (Watson et al. 1997). The predicted reduction in growing season rainfall would almost certainly affect grassland distribution, diversity, and productivity, but the effect of redistributing rainfall into fewer, larger events might be just as important (Hall and Scurlock 1991). However, the effects of the temporal distribution of rainfall events have rarely been examined (Georgiadis et al. 1989), despite the possible consequences for the conservation and sustainable use of grasslands. For example, changes in the temporal distribution of rainfall may cause the ranges of species or assemblages of conservation concern to expand, contract, or migrate. This would complicate the process of choosing lands for preservation, and managers at existing preserves may find themselves stewarding different assemblages than originally intended. A rainfall redistribution might also render ecosystems more susceptible to invasions by exotic species (Dukes and Mooney 1999). In grasslands used for agriculture, altered rainfall patterns could reduce the aboveground net primary productivity (ANPP) and shift the species composition of grasslands, affecting their carrying capacity for livestock production (Gregory et al. 1999).

In this chapter, we review the range of ecosystem responses that might result from altered rainfall regimes in mesic tallgrass prairie. We also describe preliminary results from an ongoing field experiment on the impacts of alterations in rainfall timing and amount, at the Konza Prairie Biological Station. Konza Prairie is in the Flint Hills (39°05′ N, 96°35′ W), a 1.6 million ha region spanning eastern Kansas, from the Nebraska border south into northeastern Oklahoma. This region is the largest remaining tract of unbroken tallgrass prairie in North America (Samson and Knopf 1994). Konza's climate falls within well-recognized temperature and rainfall parameters for grassland biomes. Total rainfall averages 835 mm y^{-1}, with 75 percent falling during the growing season months of April through October. Growing season rainfall is bimodal, with high monthly rainfall totals during May and June, low rainfall and high temperatures in July and August, and a second rainy period in September. High variability is common in yearly rainfall totals and in seasonal distribution (Hayden 1998).

Mesic grassland plant communities such as those at Konza Prairie are typically composed of species from several functional groups (Körner 1994). These include warm-season C_4 grasses, cool-season C_3 graminoids (grasses and sedges), and a diverse array of other C_3 herbaceous dicots (hereafter referred to as "forbs"), including nitrogen-fixing leguminous species, trees, and shrubs. Relatively pristine, frequently burned tallgrass prairies are usually dominated by two groups, C_4 grasses and C_3 forbs, and species abundance patterns in these prairies typically fit a core-satellite model (Collins and Glenn 1991; Hartnett and Fay 1998). The C_4 grasses consist of relatively few species but account for roughly 80 percent of the biomass and canopy cover (Briggs and Knapp 1995; Knapp and Medina 1999). Conversely, forbs constitute a small fraction of the biomass but a large fraction of the species.

Background

Vegetation/Rainfall Relationships

Historic changes in climatic patterns in the Central Plains have always been accompanied by changes in vegetation. Paleobotanical evidence indicates that the Central Plains have become more arid since the middle Miocene, a trend marked by the replacement of semi-open forests by grasslands (Axelrod 1985). More recently, rapid compositional changes in plant communities were witnessed during the 1930s Dustbowl era in the Central Plains (Weaver 1968). During this severe drought, which lasted from 1934 to 1941, tallgrass prairie dominants such as big bluestem *(Andropogon gerardii),* switchgrass *(Panicum virgatum),* and Indian grass *(Sorghastrum nutans)* were replaced by mid- and shortgrass species such as buffalograss *(Buchloe dactyloides),* blue grama *(Bouteloua gracilis),* and sideoats grama *(B. curtipendula).*

Modern regional patterns of grassland plant species composition and productivity also correlate with patterns of rainfall. Diamond and Smeins (1988) found a continual replacement of little bluestem *(Schizachyrium scoparium)* by big bluestem on a south-to-north temperature and rainfall gradient from coastal Texas to North Dakota. The Central Plains is a vast west-to-east gradient in grassland species composition, with the transition from shortgrass to midgrass to tallgrass prairie (Küchler 1964) following the eastward increase in annual rainfall. ANPP also increases along this west-to-east gradient (Sala et al. 1988). Relationships between ANPP and quantity of rainfall are strongest in the drier western portions of the Great Plains (Epstein et al. 1997), and weaker in the more mesic eastern portions of the Great Plains, because of the variable rainfall regime. Because Konza Prairie is located in the transition zone from mesic tallgrass to more xeric midgrass prairie and has inherently variable climatic patterns, it is well suited for studies on the interplay between rainfall quantity and timing.

Mechanisms of Mesic Grassland Responses to Rainfall

Long-term responses of grasslands to variation in the amount and timing of rainfall events should depend on three basic factors: (1) the nature of short-term, rainfall-induced dynamics in soil moisture and plant-available water, (2) long-term trends in nutrient availability, and (3) resultant plant growth and physiological responses. Soils act like capacitors, causing water that comes in relatively brief rainfall events lasting minutes or hours to remain available to plants and soil microorganisms for days to a few weeks. Plant water status responds rapidly to changes in plant-available soil moisture, and plant growth and photosynthetic carbon gain are reduced by water stress. However, differences in morphology and physiology between forb and C_4 functional groups confer different degrees of drought tolerance. For example, compared to forbs, C_4 grasses generally have higher water and nitrogen use efficiencies, photosynthetic rates, and optimal temperatures. C_4 grasses also typically possess several mechanisms of drought tolerance including leaf rolling, translocation of nitrogen to belowground storage, and tolerance of low water potentials (Knapp and Medina 1999). Leaf water potentials as low as -6.0 MPa have been recorded for the C_4 grass *Andropogon gerardii* under midsummer drought (Hake et al. 1984; Knapp 1984). Although C_4 grasses produce deep roots, most of the root mass is relatively shallow and fibrous, making C_4 grasses best suited to exploiting water high in the soil profile. On the other hand, forbs often have a greater proportion of deeper roots compared with grasses and, thus, can access deeper water supplies. Some forb species possess well-developed belowground water storage organs, which enable them to delay or avoid experiencing severe water deficits (Knapp and Fahnestock 1990). Certainly, species within functional groups vary in these traits and will vary in response to altered rainfall regimes, but functional group generalizations provide a framework for predicting general trends in diversity and productivity under altered rainfall regimes (Körner 1994).

Conceptual Model and Hypothesized Responses

Key processes affecting grassland composition and productivity responses to an altered rainfall regime are summarized in figure 9.1. This conceptual model assumes that rainfall patterns drive soil moisture dynamics. These in turn act on plant growth and physiological parameters, and on soil processes affecting nutrient availability, culminating in changes in ANPP and species diversity through time.

Under current rainfall regimes in Flint Hills grasslands, relatively fre-

Figure 9.1. Rainfall manipulation plot study at the Konza Prairie Research Natural Area in northeastern Kansas.

quent rainfall inputs keep soil moisture relatively high, resulting in only sporadic periods of significant soil water deficit. Under a rainfall scenario of fewer, larger rainfall events, soil moisture would be more likely to cycle from saturation after infrequent but heavy rainfall inputs, to severe deficit during lengthy dry periods, to saturation again with the next large event. Plant physiological status (net photosynthesis, leaf water potential) and growth (aboveground and belowground production) would vary strongly with soil moisture patterns. Consequently, ANPP is predicted to decline, because extreme soil moisture variability would cause soils to be either too wet or too dry during critical periods for growth. Over time, these declines in ANPP might be reinforced by lowered average nitrogen availability because of reduced litter inputs and nitrogen mineralization.

In the long-term, species shifts might be expected to occur, as those species better adapted to the more severe rainfall regime would assume dominance, and the community would stabilize at a new composition with lower diversity, typical of present midgrass assemblages. This is a straightforward prediction to make if total rainfall quantity is reduced to that of the present midgrass prairie. The more fundamental question is whether similar vegetation changes can be driven simply by lengthened dry intervals between rainfall inputs. The answer to this question is less clear for native vegetation assemblages than for row-crop systems, where it is well known that an auspiciously timed rainfall can significantly boost production (Wittwer 1995). On one hand, rainfall amount would be the more critical factor in composition shifts if the generally diverse grassland community buffered the effects of variation in rainfall timing, because the assemblage would likely contain species that could take advantage of rainfall whenever it falls. Conversely, since the tallgrass prairies are dominated by a relatively small number of species (the C_4 grasses), which competitively exclude many other (largely C_3) species, rainfall properly timed for C_4 grasses should maintain their dominance. Mistimed rainfall, on the other hand, should reduce their dominance, and we would expect increased abundance of previously excluded species due primarily to altered timing of rainfall events. Changes in relative proportions of species among functional groups would then be likely to affect long-term patterns in productivity and compositional change (Tilman et al. 1997; Peterson et al. 1998).

Mesic Grassland Rainfall Manipulation Experiment

Tests of these hypotheses are perhaps best accomplished by field-scale experimental manipulation of rainfall patterns in intact native communities. To this end, the Rainfall Manipulation Plot (RaMP) study was initiated at Konza Prairie. Twelve rainfall manipulation shelters were constructed in 1998 on an annually burned site with a typical mix of warm-season C_4 grasses and C_3 forbs. Our objective was to conduct a 10-year field test of the relative importance of reduced growing season rainfall amount versus increased inter-rainfall dry periods as drivers of grassland ecosystem responses to rainfall.

Each rainfall manipulation shelter consists of a galvanized tubular steel frame, a clear plastic roof, gutters, 4,200 L polyethylene tanks for rainfall collection and storage, and an irrigation system for rainfall reapplication (fig. 9.2). The shelter sides and ends are open, and the tops are covered during the growing season (May–September) with 6 mil UV-transparent polyethylene. This arrangement excludes natural rainfall from a 9 by 14 m (128 m²) area of undisturbed native tallgrass prairie, while maximizing air movement and minimizing temperature, relative humidity, and light artifacts. A 6 by 6 m area centered under each shelter is used to measure plant and soil responses to altered rainfall regimes. The core sampling area is surrounded by a 7.6 by 7.6 m perimeter barrier of 1.2 m deep galvanized sheet metal, which minimizes surface and subsoil water flow and root/rhizome penetration from outside the plot. The design and operation of these shelters is detailed further in Fay et al. (2000).

Experimental rainfall applications are conducted using a factorial combination of two treatments (n = 3 shelters per treatment): (a) altered growing season (May–September) rainfall amount, and (b) altered timing of growing season rainfall events, as follows:

1. Natural inter-rainfall dry interval and rainfall amount. This is the control treatment, which replicates the naturally occurring rainfall regime. Each time a natural rainfall event occurs, the amount of rain that fell is immediately applied to the plots.
2. Lengthened dry interval. Instead of applying rainfall immediately as it occurs, rainfall is withheld and accumulated to lengthen the dry interval by 50 percent. The accumulated rainfall is then applied as a single large event at the end of the dry interval. Over the season, the entire naturally occurring quantity of rainfall is applied, only the timing of rainfall inputs is altered.
3. Reduced amount. In this treatment, rainfall quantity is reduced by 30 percent, but is applied each time there is a natural rainfall. This imposes reduced amounts of rainfall without altering the timing of rainfall events.
4. Reduced amount and lengthened interval. Dry intervals are lengthened by 50 percent, and application quantity is reduced by 30 percent, imposing both reduced amounts and lengthened dry intervals.

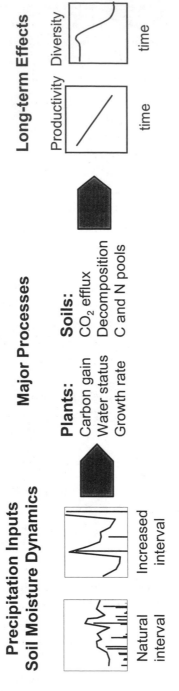

Figure 9.2. Conceptual model of the linkages between rainfall inputs, soil moisture dynamics, and plant, community, and ecosystem responses to lengthened inter-rainfall dry intervals in mesic tallgrass prairie.

These specific alterations of rainfall patterns were chosen because they are rare in current climatic patterns, but are predicted to become more common under some climate change scenarios for the Great Plains (Waggoner 1989; Easterling 1990; Houghton et al. 1990, 1996). A 30 percent reduction exceeds natural interannual variability in growing season rainfall at the RaMPs site (SD = 25 percent of the mean-based, 100-year record) and is typical of rainfall amounts presently occurring in midgrass prairies west of our study site. We based our rainfall manipulations on current rainfall patterns, rather than long-term average patterns, because year-to-year variability in rainfall is a dominant climatic characteristic of this grassland (Knapp, Briggs, et al. 1998a), making the study more realistic than if long-term averages were implemented.

In this chapter we will confine our examination of rainfall/soil moisture/plant responses to the natural interval/amount and the lengthened dry interval treatments (lengthened interval, lengthened interval/reduced amount). We chose these treatments in order to focus primarily on the impacts of altered timing, and because we expect them to exhibit the greatest impacts compared to the natural rainfall regime. Various climate, plant, and soil parameters are measured in the RaMPs experiment (Fay et al. 2000), but here we will focus on soil moisture dynamics (measured in the upper 30 cm of soil using time domain reflectometry methods; Topp et al. 1980), CO_2 flux, leaf water potential and photosynthesis, plant growth, and ANPP. Plant growth and physiological measurements focused on two species, the C_4 grass *Andropogon gerardii,* and a C_3 forb, *Solidago canadensis,* abundant members of the two dominant functional groups in the tallgrass prairie ecosystem.

The response variables that are most responsive to changes in soil moisture were measured weekly (soil moisture, soil CO_2 fluxes) or biweekly (leaf water potential and photosynthesis) from June through September. Growth of *A. gerardii* and *S. canadensis* was characterized with weekly measurements of plant height and leaf mass as plants approached peak growth (late July through September). ANPP was estimated from the dry weights of early November samples of current year standing crop. Data were summarized by calculating plot-level responses (since individual RaMPs are the experimental units), which were then used to calculate either date or growing season treatment means. All responses were measured on at least five samples per RaMP. Details of sampling techniques are in Knapp et al. (1993).

Grassland Responses to Rainfall

Rainfall and Soil Moisture Patterns

Total growing season rainfall for 1998 was about average, at 622 mm. From June through September, 27 natural rainfall events occurred (fig. 9.3), averaging 19.08 ± 3.82 mm (mean ± SE) of rainfall per event, with a rain event of at least 5 mm occurring every 8.9 ± 1.7 days. The natural rainfall regime included several large (~40 mm) storms, some occurring over several consecutive days in late June, late July, and mid- and late-September. This natural rainfall regime translated into five lengthened dry intervals averaging 25.3 ± 5.3 days, with precipitation applications following each dry period averaging 99.8 ± 34.9 mm. This nearly three-fold experimental increase in dry period length exceeded the target increase of 50 percent, because dry periods were defined using the most recent dry interval and rain events of at least 5 mm, and many of the shorter naturally occurring dry intervals also involved amounts <5 mm.

Lengthened dry intervals caused two main effects on soil moisture patterns (fig. 9.3). First, cycles of soil wetting and drying were uncoupled from the cycles associated with natural rainfall intervals. For example, during most of June, late July, and the latter half of September, naturally occurring peaks in soil moisture coincided with periods of low soil moisture caused by the experimentally lengthened dry intervals. The lengthened intervals also altered trajectories of soil moisture depletion. Soil water content fell rapidly during lengthened dry intervals in early June and mid- through late August. Rain applied at naturally occurring intervals during those same periods minimized or reversed soil moisture depletion, even with relatively small quantities (~20 mm). The lengthened dry interval/reduced amount treatment exhibited more extreme soil moisture depletion compared to the lengthened dry interval/natural quantity treatment (fig. 9.3). The lowest soil water content values (around 16 percent) were observed in the lengthened interval/reduced amount treatment in July and September after extended periods of meager rainfall. The lengthened dry interval/reduced amount treatment also reached slightly lower maximum soil water content values (43 percent) compared with the treatments receiving natural rainfall amounts (46 percent; fig. 9.3).

The ranges of dry period length and rainfall amount that exerted the most influence on soil moisture patterns during 1998 could be estimated from nonlinear regressions fit to plots of soil water content versus dry period length and rainfall amount (fig. 9.4a, b), composited from the control, lengthened

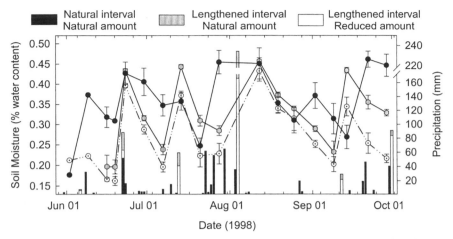

Figure 9.3. Soil moisture/rainfall relationships in native tallgrass prairie study plots during the 1998 growing season. See text for description of experimental treatments. Soil water contents (means ± SE, lines) were based on weekly time domain reflectometry measurements with 30 cm soil probes. Bars indicate rainfall applications. Figure redrawn from Fay et al. (2000).

interval, and lengthened interval/reduced amount treatments. Most soil drying occurred in the first 15 days without rainfall (fig. 9.4a), with slower soil water content reductions during longer dry periods, down to an estimated minimum of about 17.5 percent. This soil water content value is likely to represent depletion of most of the soil moisture that is readily available to plants (J. K. Koelliker, Kansas State University, pers. comm., August 1995). Soil water content increased rapidly with rainfall amount (fig. 9.4b) up to about 35 mm, suggesting that larger applications saturate the upper soil layers. This is consistent with our observations of runoff during very large applications, even when applied over two or three consecutive days to maximize infiltration (J. Carlisle, pers. obs.).

When averaged over the entire growing season, lengthened dry intervals, with no reduction in rainfall amount, caused a 9 percent reduction in soil water content compared to the rainfall applied in natural intervals and amounts (table 9.1). The lengthened dry interval/reduced amount treatment caused an average 25 percent reduction in soil water content.

Plant Responses

The reduced average soil moisture caused by the lengthened dry interval treatment was accompanied by increased plant stress and reduced growth

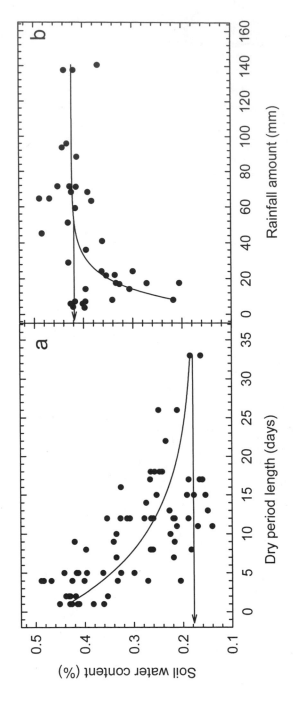

Figure 9.4. Nonlinear regressions of soil water content (swc) versus (a) days since a 5 mm or larger rain event or (b) rain event amount. Regression equations: (a) swc = 17.52 + 28.8$e^{-10.46(days)}$, $F_{2,70}$ = 46.76, R^2 = .572, $P < .0001$; (b) swc = 42.25(1−$e^{-.872(mm)}$), $F_{1,30}$ = 31.25, R^2 = .510, $P < .0001$. Arrows indicate estimated maximum/minimum swc during 1998. Data derived from figure 9.2 and Fay et al. (2000).

TABLE 9.1.

Soil, plant, and aboveground net primary productivity (ANPP) responses to rainfall manipulations, averaged (mean ± 1 SE) over the 1998 growing season. F and P values from Analysis of Variance. Letters denote significant differences between means, by least significant difference.

Rainfall Interval Rainfall Amount	Natural Natural (100%)	Lengthened Natural (100%)	Lengthened Reduced (70%)	F (df)	P-value
% Soil water content	35.51 ± 1.85a	32.65 ± 2.15a	26.78 ± 1.90b	10.89 (2,32)	.0002
Soil CO_2 production (μmol m^{-2} s^{-1})	10.10 ± 0.86a	8.59 ± 0.89b	8.46 ± 0.80b	6.59 (2,27)	.0047
Andropogon gerardii (C$_4$ grass)					
Photosynthesis (μmol CO_2 m^{-2} s^{-1})	18.12 ± 1.32a	16.35 ± 1.95b	16.16 ± 1.61b	3.40 (2,18)	.0561
Stomatal conductance (mmol m^{-2} s^{-1})	273.11 ± 39.51	229.11 ± 48.09	236.79 ± 41.16	2.12 (2,18)	.1493
Leaf water potential (MPa)	-1.69 ± 0.07a	-1.83 ± 0.07b	-1.81 ± 0.06b	10.40 (2,25)	.0005
Plant height (cm)	73.26 ± 1.77a	65.57 ± 1.27b	61.01 ± 1.36	47.97 (2,16)	.0001
Specific leaf area (cm^2 g^{-1})	108.10 ± 10.46	111.22 ± 8.62	104.06 ± 8.18	2.89 (2,43)	.0666
Solidago canadensis (C$_3$ forb)					
Photosynthesis (μmol CO_2 m^{-2} s^{-1})	14.26 ± 0.92a	12.08 ± 1.09b	12.53 ± 1.23b	16.50 (2,16)	.0001
Stomatal conductance (mmol m^{-2} s^{-1})	749.51 ± 133.98a	578.48 ± 113.49b	520.74 ± 90.80b	9.74 (2,16)	.0017
Leaf water potential (MPa)	-1.25 ± 0.06	-1.31 ± 0.06	-1.27 ± 0.06	0.42 (2,25)	ns
Plant height (cm)	56.82 ± 1.79	55.36 ± 1.85	53.41 ± 1.24	1.22 (2,8)	ns
Leaf number	56.44 ± 4.80	57.86 ± 5.48	56.05 ± 4.63	1.02 (2,8)	ns
Specific leaf area (cm^2 g^{-1})	92.51 ± 2.61a	75.34 ± 1.98b	83.21 ± 1.76b	19.01 (2,43)	.0001
ANPP (g m^{-2})	639.80 ± 38.25	574.04 ± 14.79	536.66 ± 22.12	3.72 (2,8)	.0890

(table 9.1). For the dominant warm season C_4 grass *Andropogon gerardii,* photosynthetic carbon gain, leaf water potential, and plant height were reduced 8 to 11 percent compared with plants experiencing the natural rainfall regime. The C_3 forb *Solidago canadensis* experienced reductions of 8 to 15 percent in carbon gain and specific leaf area. At the system level, the cumulative effect of lengthened dry intervals was a 10 percent reduction in both ANPP and soil CO_2 flux.

The additional soil moisture deficit caused by the lengthened dry interval/reduced amount treatment caused further reductions in plant height of *Andropogon gerardii* (8 percent) and ANPP (12 percent), but no further decreases in other plant species responses or soil CO_2 flux.

Synthesis

The initial results from the RaMP experiment verified the importance of the primary elements of our conceptual model (fig. 9.1). Changes in the timing of rainfall inputs caused altered soil moisture patterns, increased plant stress, and reduced plant growth and ANPP. These plant and soil responses suggest that changes in the timing of rainfall events will play an important role in ecosystem responses to climate change.

Increased numbers of large rainfall events separated by extended dry intervals caused strong soil moisture cycles and reduced soil moisture when averaged over the season. This experimentally induced rainfall pattern appears to be rare in the recent climatic record. For example, in a recent 15-year period (1984–1997) at Konza Prairie, only 8 percent of rainfall events 5 mm were 15 or more days apart, and only 12 percent of storms were 35 mm or larger, the threshold for soil saturation from a single rainfall event. The responses observed so far in growth, physiology, and ANPP suggest several hypotheses about long-term responses to lengthened dry intervals.

Variability in ANPP

Long-term ANPP records from Konza Prairie suggest that climatic conditions during June and July may be critical in determining annual ANPP (P. Fay, unpublished data). With more strongly cyclical soil moisture patterns resulting from a regime of lengthened dry intervals and subsequent large rainfall events, the probability is increased that soils will be either too wet or too dry during such critical times for optimum production. Thus, year-to-year variation in ANPP is hypothesized to increase. However, the lower average soil moisture suggests greater overall water limitation of ANPP. Thus, we might expect stronger interannual rainfall/ANPP correlations than have been observed previously (Briggs and Knapp 1995). Considerable spatial variability in the rainfall/ANPP relationship also has been found in prairies in the Flint Hills (Briggs and Knapp 1995), with stronger correlations on shallow-soil upland sites than on deeper-soil lowland sites. Pronounced soil moisture cycles may also lead to a stronger ANPP/rainfall relationship on lowland sites like the RaMPs site, thus reducing spatial variability in ANPP.

Reduced Dominance by C_4 Grasses

Andropogon gerardii, a C_4 grass, was more negatively affected by lengthened dry intervals than was the C_3 forb, *Solidago canadensis. A. gerardii* plants were smaller and more stressed in lengthened dry interval treatments relative to controls. In contrast, *S. canadensis* growth was not reduced by lengthened dry intervals, despite reduced leaf-level carbon gain. The stress imposed on *Solidago* by lengthened dry intervals may have been partially offset by reductions in stomatal conductance. Since grasses are typically more shallow-rooted than forbs (Weaver 1958), they may be more sensitive to soil moisture changes in upper soil horizons. These results support the hypothesis that physiological acclimation in response to climate change may delay or reduce shifts in productivity or composition (Schimel 1993).

If the responses observed in *A. gerardii* and *S. canadensis* generally hold true for C_4 grasses and C_3 forbs, there may be a trend toward reduced competitive dominance by warm season grasses. This could have several possible consequences, such as increased forb contribution to ANPP, more rapid species turnover, increased susceptibility to invasions by exotics (Dukes and Mooney 1999; Smith and Knapp 1999), or flushes of annual forb production when large rainfall inputs coincide with favorable germination conditions.

Reduced Decomposition and Nutrient Availability

The primary sources of soil CO_2, root and microbial activity, are strongly responsive to soil moisture variation (Hayes and Seastedt 1987; Knapp, Briggs, et al. 1998; Rice et al. 1998a). The sensitivity of soil microbes to water deficits suggests possible long-term effects on nitrogen availability due to changes in rainfall timing, since soil microbes account for most nitrogen mineralization (Rice et al. 1998). Year-to-year variation in the severity of July moisture deficits may induce increased variation in nutrient supply, because July is a peak period for soil microbial activity (based on seasonal CO_2 flux patterns;

Knapp, Conard, and Blair 1998). Variability in soil microbial activity could either exacerbate or ameliorate increased yearly variation in ANPP (see above).

Future Research

Rainfall Amount

Reduced rainfall amount appears to have weaker effects than lengthened dry intervals on soil moisture, plant performance, and ANPP (Fay et al. 2000). Although we would expect reduced rainfall amounts to cause plant and soil responses in our experiment, a weak quantity effect will require more years to adequately assess compared to stronger rainfall timing effects. Also, since our treatments are based on current rainfall amounts and patterns, rather than long-term averages, there should be considerable year-to-year variation in the effects of a 30 percent reduction in rainfall amount on plant and soil characteristics, with greater effects in drier years. Soil depth and texture would also influence the effects of reductions in rainfall quantity (Fredeen et al. 1997). The RaMPs study is on deep and rather clayey soils; thus, reduced rainfall could have much stronger impacts on sites with shallower or sandier soils. On the other hand, GCM model predictions regarding growing season rainfall amount are less robust than predictions for temperature (Schneider 1993; Giorgi et al. 1994; Wittwer 1995; Karl et al. 1996). So, if rainfall amount continues to exert only minor influences on ecosystem characteristics compared with rainfall timing, then GCM uncertainties regarding specific changes in rainfall amounts will become a secondary issue, and the accuracy of GCM predictions regarding increased occurrence of convective rainfall events will be a greater concern.

Offsetting Effects of Elevated CO_2 on Changing Rainfall Regimes?

In future climates, predicted changes in rainfall patterns will not occur in isolation, but in concert with increases in atmospheric CO_2 concentrations. There are a variety of predicted effects of elevated CO_2 on ecosystems, including increased plant productivity, water use efficiency, soil moisture, and soil carbon storage, as well as decreased decomposition and nitrogen mineralization (Parton et al. 1995; Coughenour and Parton 1996; Mooney et al. 1999). For the Central Plains grasslands, altered rainfall patterns could potentially offset some CO_2 effects, because our preliminary results indicate that a lengthening of dry intervals is likely to decrease soil moisture and productivity. Thus, an important next step in unraveling the impacts of climate change on the conservation and sustainable use of grassland ecosystems would be a field-scale rainfall-times-CO_2 experiment, where the potential for rainfall patterns and elevated CO_2 to offset each other can be rigorously tested.

Climatic patterns, along with ungulate grazing and fire, are the three primary factors governing the structure and function of grassland ecosystems. Even in its early stages, the Rainfall Manipulation Plot experiment has provided useful insights into the impacts of altered precipitation patterns, and increased soil moisture variability, on fundamental characteristics of tallgrass prairie, an important temperate grassland ecosystem. The knowledge gained from this experiment over a long-term period will sharpen our understanding of the climatic context in which fire and grazing impacts occur. In addition, we expect to gain a greater appreciation of how precipitation variability may interact with other climatic and nonclimatic elements of global biological change.

Acknowledgments

The Rainfall Manipulation Plot study is supported by USDA NRICGP (96-00713), NIGEC, and NSF (96-03118), the Konza Prairie LTER program (http://www.konza.ksu.edu), and the Kansas Agricultural Experiment Station. We thank Roger Baldwin, Scott Heeke, Michelle Lett, Josie Pike, and the staff of the Konza Prairie Research Natural Area, Tom Van Slyke, Jim Larkins, Dennis Mossman, Ryan Anderson, and Chris Holliday for their contributions to this research. We also thank Jim Koelliker, Department of Biological and Agricultural Engineering at Kansas State University for his insight into tallgrass prairie soil moisture dynamics. Konza Prairie is a preserve of The Nature Conservancy, managed for ecological research by the Division of Biology, Kansas State University. Publication number 00-209-B of the Kansas Agricultural Experiment Station.

10

Responses of Eastern Deciduous Forests to Precipitation Change

PAUL J. HANSON, DONALD E. TODD, DALE W. JOHNSON,

JOHN D. JOSLIN, JR., & ELIZABETH G. O'NEILL

Changes in average global land surface temperatures during the next century will likely modify regional and global hydrologic cycles, leading to increased winter precipitation at high latitudes, more extremely hot days and fewer extremely cold days, and changes in the number of droughts or floods depending on location (Kattenberg et al. 1996; Rind et al. 1990; Houghton et al. 2001; see also chapter 1, this volume). These predicted changes in climate are cause for concern when considering productivity of terrestrial ecosystems, biogeochemical cycles, and the availability of water resources (Kirschbaum and Fischlin 1996; Melillo et al. 1996). Concerns regarding vegetation impacts are amplified because change is expected to occur much faster than past forest succession processes and tree seed dispersal rates (Davis 1989; Overpeck et al. 1991; Pastor and Post 1988).

The direction and magnitude of future changes in precipitation remain uncertain, and recent scenarios for regional climatic change suggest a wide range of future climates and precipitation scenarios (VEMAP Members 1995; Schimel et al. 2000). Manipulative field experiments can be used to evaluate the level and magnitude of ecosystem responses to a variety of precipitation change scenarios.

To study impacts of potential precipitation change on forest productivity and ecosystem processes, Hanson et al. (1995, 1998) implemented the Walker Branch Throughfall Displacement Experiment (TDE) in 1993. The TDE is a large-scale field experiment designed to modify throughfall amount in an upland oak forest over multiple years. The TDE is located in the humid region east of the Mississippi River and is representative of the Eastern Broadleaf Forest Province as defined by Bailey et al. (1994). Droughts in this region are limited in duration and spatial extent, and tend to occur late in the growing season. Years without any drought are also possible (Hanson and Weltzin 2000). The seasonality, intensity, and duration of the droughts that do occur are not predictable using current numerical weather prediction models. This chapter summarizes the TDE experiment and describes key conclusions based on data from six years of sustained manipulation.

Methods

The experiment was located on the Walker Branch Watershed (35°58' N, 84°17' W), a part of the U.S. Department of Energy's (DOE's) National Environmental Research Park near Oak Ridge, Tennessee (Johnson and van Hook 1989). Long-term (50-year) mean annual precipitation was 1352 mm and mean annual temperature was 14.2°C. The acidic forest soils (pH 3.5 to 4.6) are primarily typic Paleudults. Depth to bedrock at this location is approximately 30 m. The site was chosen because of its uniform slope, consistent soils, and a reasonably uniform distribution of vegetation. *Quercus* spp. and *Acer* spp. were the major canopy dominants, *Liriodendron tulipifera* L. was a canopy dominant on the lower slope positions, and *Nyssa sylvatica* Marsh. and *Oxydendrum arboreum* (L.) D. C. were the predominant species occupying midcanopy locations. In March 1994, stand basal area averaged 21 m^2 ha^{-1} across the TDE site with nearly identical basal area on each plot. The number of saplings (trees <0.1 m diameter at breast height [dbh]) across the TDE area averaged 3073 ha^{-1} in 1994. *Acer rubrum* L. and *Cornus florida* L. combined to make up 59 percent of all saplings and 53 percent of the sapling basal area.

Experimental System

The experimental system and its performance were described in detail by Hanson et al. (1995, 1998), and important methodological issues relevant to large-scale water manipulations are detailed in Hanson (2000). The manipulations of throughfall reaching the forest floor were achieved by gravity-driven transfers of precipitation throughfall from "dry" (-33 percent) to "wet" (+33 percent) treatment areas. There were three 80 by 80 m plots in the TDE: one wet, one dry, and one ambient. Each plot was divided into 100 subplots (8 by 8 m) for repetitive, nondestructive measurements of soil and plant characteristics. On the dry plot, throughfall precipitation was intercepted in ~1,900 troughs (0.3 by 5.0 m) made of greenhouse grade polyethylene. The troughs were suspended at an angle above the forest floor of the dry plot and covered ~33 percent of the ground area. The intercepted throughfall was transferred across the ambient plot and distributed onto the wet treatment plot through paired drip holes spaced approximately 1 m apart. The troughs were arranged in 21 rows of ~80 to 90 troughs each. To evaluate impacts of the trough infrastructure on understory conditions, subplots of troughs containing holes were included on the wet and ambient plots during the first three years of the study. The subplots were removed in 1996 after the effects of troughs on understory microclimate were found to be small (Hanson et al. 1995).

The experimental area was located at the upper divide of the watershed, so that lateral flow of water into the plots from upper slope positions did not occur. The site had a southern aspect, which enhanced radiation exposure and generated greater soil moisture differentials between treatment plots. Reductions in soil moisture anticipated from the experimental removal of 33 percent of the throughfall were designed to be comparable to the driest growing season of the 1980s drought (Cook et al. 1988).

Soil Water Content, Water Potential and Weather Measurements

Soil water content (percent, v/v) was measured with a time domain reflectometer (TDR; Soil Moisture Equipment Corp., Santa Barbara, California) following the procedure of Topp and Davis (1985) as documented for soils with high coarse fraction content (Drungil et al. 1987). Sampling locations were installed at an 8 by 8 m spacing across the TDE site, giving more than 100 soil water monitoring locations per plot. At each location two pairs of TDR waveguides were installed in a vertical orientation (0–0.35 and 0–0.7 m). The TDR measurements were adjusted for the coarse fraction of these soils (mean coarse fraction of 14 percent) and converted to soil water potentials (SWPs) using soil moisture retention curves for the A, A/E, and E/B horizons of these cherty silt loam typic Paleudult soils (Hanson et al. 1998).

To facilitate comparisons of soil water deficit severity between years, we report the minimum SWP (MPa) and calculate a water stress integral (units of MPa d) for all years and treatments. Weather data including air temperature, relative humidity, and soil temperatures (at 0.1 and 0.35 m) were logged hourly on each treatment plot. Rainfall, solar irradiance (Pyranometer sensor, LiCor Inc., Lincoln, Nebraska) and photosynthetic photon flux density (Quantum sensor, LiCor Inc.) were also measured continuously and logged as hourly means. From January 1997 to February 1998 hourly litter temperatures were recorded in four locations each in the dry and ambient plots using miniaturized data loggers (Onset Computer Corporation, Bourne, Massachusetts).

Growth Measurements

One hundred seventy selected mature trees 0.2 m dbh were fitted with dendrometer bands for biweekly measurements of stem circumference during each growing season (Hanson et al. 2001). The trees were selected from five species *(Quercus alba* L., *Q. prinus* L., *Acer rubrum* L., *Liriodendron tulipifera,* and *Nyssa sylvatica)* that made up almost 80 percent of the basal area of the TDE experimental area. A single dendrometer measurement consists of duplicate digital caliper measurements (0.01 mm resolution) of the distance between two reference holes in stainless steel dendrometer bands (25.4 mm wide by 0.2 mm thick).

In February and March of 1994, 10 transects were established to observe growth of saplings (plants < 0.1 m dbh) across the three plots of the TDE from lower- to upper-slope positions. Although other species were considered for these measurements, only *Acer rubrum* and *Cornus florida* were distributed across the TDE in sufficient numbers for inclusion. Starting at the time of spring leafout each year, biweekly measurements of stem diameter at a permanently marked location on each sapling's main stem (typically between 1 and 1.5 m above the ground) were made until sapling growth had ceased for that year. Each stem caliper measurement was the mean of three replicate diameter measurements made with a digital caliper (0.01 mm resolution) around the point of measurement. Replicate measurements were required to minimize the impact of noncircular stem cross sections.

Root growth was estimated by observing root elongation along the surface of buried transparent minirhizotron tubes distributed along upper and lower slope positions (Joslin and Wolfe 1998; Joslin et al. 2001). In each of the TDE treatment plots, 10 tubes were installed, 5 each along the lower and upper transects.

Foliar Production, Litter Chemistry, and Decomposition

Fallen leaves and other materials (i.e., twigs, seeds, etc.) were collected from 147 baskets (0.53 by 0.38 m) located near the central 49 (7 rows by 7 columns) grid intersections of each of the three treatment blocks of the TDE. Litter was collected periodically throughout the year (typically mid-May, late August, and biweekly from October 1 through December 1). Litter samples were dried (one to two days at 70°C) and sorted to separate foliar from nonfoliar components. Foliage litter mass was divided by the basket collection area (0.2 m^2) to yield annual foliar production on a ground area basis.

Foliar litter chemistry was measured annually, using *Q. alba* and *A. rubrum* as indicator species. Six litter collection baskets were chosen per treatment, three upslope and three downslope. Litter was subsampled from these baskets each year, combined by species, and analyzed for total carbon and nitrogen (Carlo Erba C-N analyzer, Carlo Erba Strumentazione, Milan, Italy).

A long-term reciprocal litterbag study was initiated in 1995 to examine the effects of both litter source and environment on decomposition. *Acer rubrum, Quercus prinus,* and *Cornus florida* litter was collected in the fall of 1995 from each treatment plot (hereafter referred to as "Source"). Nylon mesh bags, 25 by 25 cm, were filled with a mix of the three species to predetermined mass ratios reflecting the abundance of each species on the TDE site. Lignin to nitrogen ratios were determined for each species from each Source, as well as for the final mixtures, using near-infrared spectroscopy (Wessman et al. 1988). Twelve replicate bags from each Source were placed at random locations on each plot (hereafter referred to as "Site") just below the Oi layer. A subset of four bags from each Source by Site combination was retrieved after one, two, and five years in the field, and litter mass loss and residue carbon and nitrogen were determined. Litter from the first and second collections was ground and analyzed for standing microbial biomass using the substrate-induced respiration method (Beare et al. 1990).

Soil Chemistry, Organic Layer Carbon, and Soil Nitrogen Flux

A detailed description of the sampling of soils for available nutrients and nitrogen flux through the soil profile can be found in Johnson et al. 1998. Soil solutions were collected with ceramic cup tension lysimeters (Soil Moisture Equipment Corp., Santa Barbara, California) installed at 25 and 70 cm. The lysimeters were installed prior to treatment in a three by three array in each sub-watershed at a spacing of 7.9 m. Prior to collection of samples, tensions of 40 kPa were established in each lysimeter. In addition to the ceramic cup lysimeters, Johnson et al. (in press) installed resin lysimeters beneath the O horizons of all treatments. The resin lysimeters consisted of a 5.5 cm long, 4 cm inside diameter PVC pipe, within which a resin bag was sandwiched between layers of washed silica sand.

Soil organic layers were measured in 1992, prior to the experimental treatments beginning, and in February of 1999, as an early indication of the impacts of the throughfall manipulations on heterotrophic decomposition and nitrogen cycling. Sample locations were designated in each treatment plot along upper, middle, and lower slope transects. In each treatment area, five locations were designated per transect for a total of 15 sampling locations in each treatment area. To minimize any impact of trough placements on litter accumulation or decomposition, posttreatment sampling within the dry treatment plot included twice the area (i.e., one under and one between troughs) per location. Identical sampling locations were avoided between years. Organic horizons were sampled within 0.25 m² circular plots. Coarse twigs and large pieces of nonfoliar litter were removed first, followed by the collection of the foliar component of the Oi horizon (Oi; Soil Science Society of America 1997). The Oe and Oa horizons were sampled together (Oe/Oa; Soil Science Society of America 1997). All samples were oven-dried for two days (100°C) and dry mass was determined. Complete samples were subsequently ground in a Wiley mill to pass through a ~20 mesh screen, and a portion of the homogenized ground material was combusted at 500°C to obtain ash content. Total dry mass per unit ground area (m²) for each organic horizon was corrected for ash content.

Measured concentrations of nitrogen in soil solution and modeled estimates of water flow through the rooted soil profile (unpublished data) were used to estimate annual nitrogen losses from the soil in each treatment.

Statistical Analyses

The unreplicated nature of the TDE is not ideal, but the resulting pseudoreplication is recognized as a reasonable approach when such costly experimental field designs are undertaken (Eberhardt and Thomas 1991). Nevertheless, dealing with the issue of pseudoreplication in our sampling design is critical (Hurlbert 1984). Hanson et al. (1998) showed that the individual 8 by 8 m resolution soil water measurements across the TDE plots were not correlated with each other and could therefore be appropriately treated as independent measurements. Using similar logic, we considered spatially separated plants independent of one another with respect to soil water conditions (Hanson 2000; Hanson et al. 2001). With these considerations in mind, we used analyses of covariance to test for wet, ambient, and dry treatment effects on growth of mature trees and saplings. We used univariate analysis of variance to determine treatment effects on root growth, litter layer, and nutrient cycling response variables.

Results and Discussion

1993–1999 Climate

Weather and soil water data showed substantial interannual variation from 1993 through 1999 (fig. 10.1). Lower than average annual precipitation was measured in 1993 (-16 percent), 1995 (-16 percent), 1998 (-9 percent), and 1999 (-15 percent), and above average precipitation was observed in 1994 (+24 percent), 1996 (+21 percent), and 1997 (+8 percent). Growing season precipitation was near normal in 1994 and 1999, but was 26 to 38 percent below normal during the drought years of 1993, 1995, and 1998. Growing season precipitation was 47 and 22 percent above normal in 1996 and 1997, respectively. Mean annual air temperature and annual incident solar radiation were not as variable as annual precipitation, but 1998 and 1999 were warmer than average (fig. 10.1A,B). Surveys of air and soil temperatures and humidity near the ground across the TDE (Hanson et al. 1995) showed that there was little or no effect of the experimental infrastructure on below-canopy temperature and humidity. However, data from litter temperature surveys throughout 1997 suggested that forest floor temperatures were periodically impacted by the troughs through a combination of shading and radiant heat retention (E. G. O'Neill, unpub. data).

Annual cumulative incident solar radiation at the site was similar across years. Cumulative growing season radiation inputs ranged from 2643 to 3155 MJ m^{-2} from 1993 through 1999, but showed few trends with annual or growing season precipitation.

TDE Treatments and Performance

The TDE experimental system produced statistically significant differences in soil water content in years having both dry and wet conditions (Hanson et al. 1998; Hanson 2000). During summer drought periods, the dry plot soil water content reached lower values before those of the ambient and wet plots, and dry plot water content was the last to return to field capacity after canopy senescence in the fall. Over a typical annual cycle including a late season drought, surface soil water content (0–0.35 m) ranged from 26 to 8 percent (v/v). Maximum differentials of soil water content between wet and dry plots in the surface soil horizon were 8–10 percent during summers with abundant precipitation and 3–5 percent during drought periods (data not shown). Treatment impacts on SWP were largely restricted to the surface soil layer. Because annual or growing season precipitation inputs by themselves do not

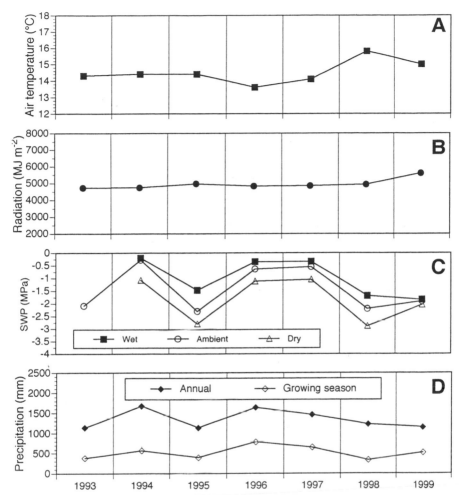

Figure 10.1. Weather and soil water variables for each year of the study, including (A) mean air temperature for the year, (B) cumulative radiation for the year, (C) minimum soil water potential (swp) by treatment and year, and (D) total annual and total growing season precipitation. The growing season was defined as May through September.

necessarily characterize the intensity of drought with respect to biological responses, we calculated two additional indices of soil moisture deficit: minimum daily swp (fig. 10.1C) and the water stress integral (table 10.1). Each index was estimated from a combination of measured swp data and modeled interpolations of the data (Hanson et al. 2001). Minimum daily swp is an instantaneous measure of maximum observed soil dryness in a given year. swp ranges from 0 (no stress) to values less than -2.5 MPa (severe stress). Water stress integral is the summation of daily mean swps throughout the year (units

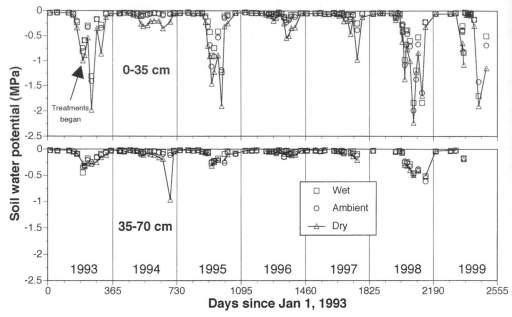

Figure 10.2. Soil water potential for the 0–0.35 m (A) and 0.35–0.7 m (B) soil depths from 1993 through 1999. Data are the mean value for the wet, ambient, and dry plots on the Walker Branch Throughfall Displacement Experiment (TDE). Throughfall displacement treatments were initiated on July 14, 1993.

of MPa days), and it represents a better characterization of the intensity of drought.

Minimum daily SWPs showed that significant drought occurred in 1993, 1995, 1998, and 1999. However, the annual water stress integrals showed that the 1993 drought was not sustained as long as those in 1995, 1998, and 1999. In the wet growing seasons of 1994, 1996, and 1997, ambient SWP never fell below -0.7 MPa, but the SWP in the dry plot reached substantially lower levels (fig. 10.2; table 10.1). Even though ambient precipitation inputs during the dry 1995 growing season were comparable to reduced dry plot precipitation inputs in 1994 (394 vs. 379 mm), the minimum ambient versus dry plot SWPs were different (-2.3 vs. -1.1 MPa, respectively) because of the timing of rainfall events. Calculated water stress integrals for the January through July versus the August through December periods further demonstrate year-to-year differences in the characteristics of droughts (table 10.1). Whereas the droughts of 1995, 1998, and 1999 were most severe (i.e., exhibited the most negative SWPs) in the second half of the growing season, the drought of 1993 was most severe during the first half.

Table 10.1.

Annual and growing season (GS) water stress integrals (MPa d) by treatment and year. The growing season was defined as May through September. Values for the integrated water stress integral are estimates from a model of the water budget for these stands.

Variable	1993	1994	1995	1996	1997	1998	1999	Mean
Ambient plot								
January/July	−61	−14	−48	−13	−11	−20	−21	−27
August/December	−11	−9	−62	−16	−15	−73	−77	−38
Annual	−72	−23	−110	−29	−26	−93	−98	−65
Dry plot								
January/July	—	−23	−72	−15	−12	−26	−40	−31
August/December	—	−10	−96	−30	−25	−149	−120	−72
Annual	—	−33	−168	−45	−37	−175	−160	−103
Wet plot								
January/July	—	−12	−27	−13	−11	−16	−14	−15
August/December	—	−9	−38	−12	−12	−31	−57	−27
Annual	—	−21	−65	−25	−23	−47	−71	−42

Plant Growth

Multiyear growth and physiological observations through six years of manipulation show that the small stature vegetation (e.g., seedlings and saplings) is the most sensitive to reductions in rainfall inputs (Hanson et al. 2001; Holmgren 1996). Mean annual sapling basal area increment throughout the six-year observation period became significantly higher in the wet plot compared to the ambient and dry plots (fig. 10.3). Conversely, no significant changes in annual mature tree basal area increment were observed across treatments. Limited rooting depth and carbohydrate supplies in saplings and seedling were suggested as mechanisms for their enhanced sensitivity (Hanson et al. 2001).

Although insensitive to the chronic TDE manipulations, mature tree growth did show significant interannual changes, with growth in some dry years attaining only 30 to 50 percent of the growth in wet years depending on the species. The droughts in 1993 and 1995 caused reduced annual basal area growth in the wet growing seasons of 1994 and 1996, but severe late-season droughts in 1998 and 1999 had little impact on this measure (data not shown; Hanson et al. 2001).

Maximum daily growth rates occurred before the end of June in each

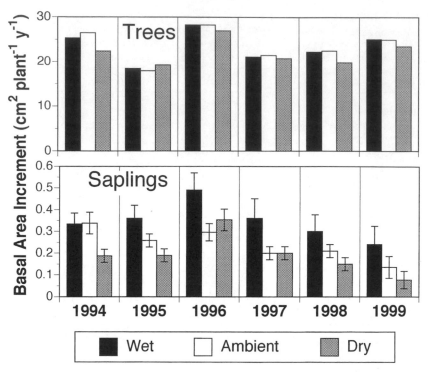

Figure 10.3. Mean annual growth ± sᴇ of mature trees (dbh > 0.1 m) and saplings (dbh < 0.1 m) by year and treatment.

year (day 180), and all basal area growth was completed before late September (day 270) in both wet and dry years (Hanson et al. 2001). No stem basal area growth was observed from late September through October, even though canopy foliage remained intact and active throughout that period (Wilson et al. 2000). Hanson et al. (2001) concluded that basal area growth in upland oak forest stands would seldom be impacted by late-season droughts, and suggested that a minimum of 60 percent of annual basal area growth would always be completed before the development of summer drought.

Root Growth

Root growth was reduced during drought periods, but it also exhibited significant increases in the dry treatment in 1996, after soil water was restored to favorable levels (Joslin and Wolfe 1998; Joslin et al. 2001). Regrowth of roots following drought periods may be a significant process whereby forest ecosystems can adapt to changing rainfall patterns in the future. Hypothesized changes in rooting density with depth due to the ᴛᴅᴇ treatments (i.e., deeper versus shallower roots on the dry or wet plots, respectively) have yet to be shown, but comprehensive measurements of rooting density with depth following six years of manipulation are underway.

Based on minirhizotron observations from 1995 through 1999, Joslin et al. (2000) have established a highly significant ($P < 0.0001$) log linear relationship between swp and root elongation index (i.e., the measure of root growth from minirhizotrons). Although their data showed that swp plays a major role in controlling root growth, they also indicated that phenologically related factors can override environmental controls. Root growth was observed to peak every year immediately following completion of leaf expansion congruent with the fixed shoot growth pattern of the mature trees (Hanson et al. 2001; Hanson and Weltzin 2000). Therefore, data for both above- and belowground growth suggests that these forests typically complete much of their annual growth before late-season droughts can have a major impact on current year productivity.

Leaf Litter Chemistry and Decomposition

Lignin and nitrogen concentrations determined in 1995 upon initiation of the litterbag study showed no effect of the dry treatment for either *Q. alba* or *A. rubrum,* although litter from the wet treatment (both species) had significantly lower nitrogen, higher lignin, and a higher lignin:nitrogen ratio than litter from the other two treatments. The effect on *A. rubrum* was more pronounced than on *Q. prinus.* Analysis of changes in leaf litter chemistry of each new annual cohort continues.

Carbon Accumulation in Litter

Through six complete years of TDE manipulations, mean annual foliar litter production has been nearly constant at 500 g dry matter per m², and annual foliar production was unexpectedly insensitive to TDE treatments. In addition, although nonfoliar litter (i.e., flowers, twigs, and seeds such as acorns) varied considerably from year-to-year, it has not been impacted by TDE treatments.

Whereas pretreatment sampling revealed no significant differences in organic horizon mass among treatment areas, six years of throughfall manipulation resulted in a significant increase in organic horizon mass on the dry treatment plot (fig. 10.4). Specifically, the Oe/Oa layer and nonfoliar litter were 57 percent and 77 percent higher, respectively, in the dry plot relative to the ambient or wet plots. When all organic layer components were combined (Oi, Oe/Oa, and coarse wood), the standing pool of organic matter was

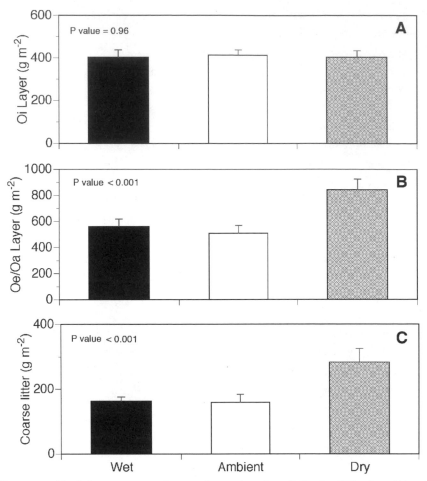

Figure 10.4. Total dry mass per unit ground area ± SE (g m⁻²) for the Oi horizon (A) and Oe/Oa horizon (B) and the nonfoliar coarse litter (C) after six years of throughfall manipulation. Data are corrected for ash content.

35 percent greater in the dry plot than in the ambient plot after the first six years of manipulation.

These results should be comparable to decomposition rates determined from the litterbag study. After one year in the field, mass loss for all Sources combined was greater in the wet and dry plots (60 and 64 percent original mass remaining, respectively) relative to the untreated plot (70 percent remaining). By the end of the second year, however, there were no differences in mass loss due to Site. If only those data that represent pairings of Source and Site (i.e., litter produced and decomposed on the same plot) are examined, then there is no difference due to altered precipitation for either year. These results compare reasonably well with the results above on organic matter accumulation. In the Oi layer, equivalent to the first and second year litterbags, no differences were seen in organic matter mass. It is only in the Oe/Oa layers that differences were seen; material in the litterbags in the field for five years correspond to this layer, and we expect to see more litter remaining in the bags from the dry plots relative to the wet or ambient.

Dry matter accumulation in the organic layer is a dynamic process driven by the balance between litter inputs (leaf, seed, branch, and fine roots) and losses resulting from microbial or fungal decomposition, invertebrate consumption, and transfer of humic materials to the mineral soil horizons. Litter mass accumulation indicates reduced decomposition and carbon accumulation, but carbon accumulation in the forest litter layer might increase future fire hazards and lead to long-term reductions in nutrient availability.

Nutrient Cycling

A combination of direct measures of soil nutrient status and nutrient cycling simulations have been used to evaluate the impact of TDE treatment scenarios on nutrient availability (Johnson et al. 1998, 2000, in press). Early results from the TDE after only four growing seasons (i.e., 1997) suggested that sustained reductions in throughfall inputs would slow the rate of soil acidification and phosphorus loss from eastern deciduous forests, but would not materially affect growth or ecosystem nitrogen status (Johnson et al. 1998). More recent data through the 1999 growing season (Johnson et al. in press) showed significant treatment effects for all measured ions except NH_4+ and NO_3- in the Bt horizon (which were at or near trace levels at all times). However, in the organic horizons, large and statistically significant treatment effects on nitrogen fluxes (measured by the resin lysimeters) were apparent (i.e., nitrogen flux was reduced in the dry plot and enhanced in the wet plot). These data suggest net uptake of nitrogen in the forest floor, which is consistent with immobilization of nitrogen in the litter, as previously described.

Nitrogen content of the organic horizons during February of 1999 ranged from 22 to 32 gN m^{-2}. The dry plot nitrogen content was ~9 gN m^{-2} higher than the ambient (data not shown), suggesting an annual rate of nitrogen immobilization in the dry plot organic horizons of 1.5 gN m^{-2} y^{-1}. This rate of immobilization is less than observed differences in leaching from the O horizons during 1999 (0.4 gN m^{-2} y^{-1}; Johnson et al. in press). It is comparable to amounts sequestered in annual wood production (Trettin et al.

1999) or atmospheric nitrogen deposition (Johnson and Lindberg 1992), but three to four times less than the mineralization rates from the A horizon (Garten et al. 1994). Immobilization of nitrogen in forest organic horizons might reduce its availability for growth and physiological function if sustained over long time periods. In addition, a buildup of organic horizon mass could also increase the potential for future ground fires, which would result in a rapid loss of significant nitrogen from the forest ecosystem.

Conclusions

Stand-level experiments were used to understand mechanisms of forest ecosystem response to changes in regional rainfall that may result from a warming global climate. Growth of mature trees appears buffered against incremental changes in precipitation, but growth of seedlings (Holmgren 1996) and saplings of certain species (e.g., *Cornus florida* and *Liriodendron tulipifera*) may be severely reduced under future droughts. This conclusion suggests that incremental changes in precipitation patterns are unlikely to have large impacts on productivity of standing timber species in eastern forests, but stand composition of deciduous forests could be altered. However, litter layer dry mass accumulation leading to reduced nitrogen availability to plants in the dry treatments suggests that changes in nutrient cycling could eventually impact growth of mature trees.

Differences in seasonal timing of rainfall will have a greater impact on plant productivity (or carbon sequestration) than changes in rainfall applied equally throughout a year. For example, early spring droughts have a larger impact on forest productivity, because they coincide with the period of maximum plant growth. A key implication of this conclusion is that accurate predictions of plant, forest stand, and ecosystem responses to changing regional climates will require a concomitant understanding of future climate dynamics. The temporal resolution of current precipitation change scenarios (i.e., more or less rainfall annually) must be improved to include scenarios of rainfall periodicity. This requirement is essential, because it is the balance between rainfall inputs and ecosystem water use that lead to quantification of soil water status. Soil water status is a key variable underlying accurate predictions about the impact of precipitation change on terrestrial ecosystems.

The TDE is slated to operate through the growing season of 2003 to complete observations of long-term cumulative mature tree and soil responses, fill out mechanistic plant/ecosystem response relationships, and support ongoing and externally funded collaborative research efforts. Final results from TDE research will yield a multicomponent picture of the response of forests to rapidly changing rainfall patterns, which is needed for societal assessments of the influence of climate change on forest productivity and carbon sequestration.

Acknowledgments
This research was sponsored by the Program for Ecosystem Research, Environmental Sciences Division, Office of Biological and Environmental Research, U.S. Department of Energy under contract DE-AC05-00OR22725 with University of Tennessee-Battelle LLC. Research was conducted on the Oak Ridge National Environmental Research Park. Publication No. 5030, Environmental Sciences Division, Oak Ridge National Laboratory. D. W. Johnson's contribution was also funded in part by the Agricultural Experiment Station, University of Nevada, Reno.

11

Assessing Response of Terrestrial Populations, Communities, and Ecosystems to Changes in Precipitation Regimes

Progress to Date and Future Directions

JAKE F. WELTZIN & GUY R. MCPHERSON

As a whole, the chapters in this volume illustrate the many complex, tightly woven, interactive relationships between precipitation, soils, and plants. Not only do these relationships occur under current environmental regimes, but they will also dictate the response of ecological systems to potential future precipitation and soil moisture regimes. At the risk of some redundancy, we attempt in this chapter to summarize and synthesize the important messages of the core chapters of the book. We then discuss potential future directions for research that can help provide answers to the many questions that remain, or that were indeed generated by the research and analyses herein.

As described by McAuliffe (see chapter 2, this volume), soil is a sieve through which precipitation influences plants. Vegetation response to precipitation is mediated by soil morphological characteristics and conditions such as texture, structure, the arrangement of pores, depth, and spatial heterogeneity. These attributes exert considerable influence over the vertical and horizontal distribution of soil moisture and, therefore, roots. For example, a shallow argillic horizon may limit downward movement of soil water, with consequent impacts on root development. In addition, soil characteristics interact with morphology, phenology, and physiology of plants. As Mary Austin recognized as early as 1903 (see foreword), this "belowground connection," reviewed by Williams and Snyder in chapter 3, imposes constraints on plants that are expressed in the aboveground abundance, distribution, and form of individual plants, and the vegetation as a whole.

Because of these strong interactions between the plant and its substrate, it is not particularly easy to predict the response of vegetation to current, let alone potential future, precipitation regimes. For example, long-lived perennials, or "extensive exploiters" (see chapter 2), may be buffered from even extended deficits in precipitation and soil moisture irrespective of soil characteristics. In contrast, "intensive exploiters" with dense, shallow root systems, such as annual grasses and dicots, are much more tightly coupled to intra- and interannual patterns of precipitation and the characteristics of their soil substrate. Thus, explaining and predicting the response of vegetation to altered precipitation regimes at multiyear and decadal scales (which are most relevant to management decisions and activities) must rely on detailed information about soils (e.g., depth, texture, proportion of rock fragments) *and* plants (e.g., characteristics of functional types, species, and ecotypes or genotypes).

Schwinning and Ehleringer (2001) further explored the relationship between water, soil, and plants using a hydraulic soil-plant model to examine trade-offs between water use and adaptations in plants adapted to pulse-driven arid ecosystems. They used a genetic algorithm to identify character suites (i.e., phenotypes) that maximize photosynthetic carbon gains for plants that experience different scenarios of spatial (shallow vs. deep) and temporal (equilibrium and drawdown) distributions of soil water. As the availability of shallow (usually summer-derived) and short-term (or "pulsed") soil water increased, optimal phenotypes exhibited characteristics that maximized use of pulse water: small root-to-shoot ratios, shallow root systems, high leaf conductance with high stomatal control. In contrast, as deeper, more constant (winter-derived) sources of soil water increased, phenotypes shifted towards adaptations that maximized deep water use: large root-to-shoot ratios, deep root systems, lower leaf conductance with low stomatal control.

The continuum of such morphological and physiological characteristics in response to spatial and temporal variations in soil water corresponds well with three groups of phenotypes observed in arid and semiarid systems: (1) winter annuals and drought-deciduous perennials that depend primarily on deeper soil water for carbon gain, (2) succulent perennials that depend exclusively on shallow pulses of soil water, and (3) evergreen perennials that exploit whichever water source is available at the time. These results support prior observations that plants in arid and semiarid environments appear to be adapted and specialized to particular spatial and temporal patterns of soil moisture (see chapters 2 and 3). Model results are in turn supported by empirical observations that changes in the timing or duration of soil moisture may determine plant species composition or production on local to regional scales (see chapters 4, and 6–10).

Biogeographic models are well suited to examine broadscale (i.e., regional to continental) relationships between precipitation and vegetation (see chapter 4). Tight correlations exist between the temporal and spatial distribution of regional air masses, which dictate climate, and the type and distribution of vegetation at the regional scale. These linkages attest to the importance of precipitation regime and the interaction of precipitation with other factors, both abiotic (e.g., soil texture, temperature, periodicity of large-scale atmospheric circulation, fire) and biotic (competition, herbivory, characteristics of individual plant species). For example, the Prairie Peninsula in the midwestern United States receives enough annual precipitation to support a closed-canopy forest, but a precipitation regime that results in wet but variable summers and typically dry winters tends to favor dominance by relatively shallow-rooted grasses (see Neilson, chapter 4, this volume).

Further, changes in the distribution of vegetation must be considered within horizontal (i.e., latitude and longitude), vertical (elevation), and temporal dimensions. Accurate simulation of these large-scale patterns will require process-based modeling approaches that capture the detailed local complexity and temporal patterns of vegetation change. In particular, newly developed dynamic global vegetation models (DGVMs) can simulate vegetation change through time in response to transient changes in climate, and can incorporate stochastic but predictable processes such as plant recruitment, migration, and dieback, as well as abiotic factors such as fire and interannual variation in climate related to global anomalies such as El Niño/La Niña events (Daly et al. 2000; Bachelet et al. 2001). For example, in a recent intermodel comparison of six DGVMs, regional-scale changes in vegetation distribution, structure, and production were predicted to occur in response to surface warming, especially when coupled with shifts in regional precipitation regimes (Cramer et al. 2001).

On smaller scales, direct manipulation of precipitation is a robust strategy for investigating mechanisms of interactions between soil, soil moisture, and plants (see chapters 6–10). Field-oriented research with in situ manipulation or simulation of precipitation can be used to determine how these factors may respond individually and interactively to current and future soil moisture regimes. Importantly, field studies can be conducted at temporal and spatial scales most relevant to the plant population, community, or ecosystem of interest.

There are several approaches to manipulation of soil moisture in the field (e.g., irrigation, supplementation of ambient precipitation using sprinklers or snow fences, trenching or terracing). However, because temporal and spatial variations in precipitation are the norm rather than the exception, the tightest control over experimental, designated soil moisture levels can be realized with the use of precipitation shelters. As discussed by Owens (see chapter 5, this volume), precipitation shelters can vary in size depending on the system in question, and can utilize any number of approaches designed to maintain target precipitation regimes and soil moistures. The design, construction, and maintenance of a replicated, dependable set of precipitation shelters requires considerable planning and expense; decisions about the appropriate shelter system for any experiment will ultimately encounter economic and logistical trade-offs vis-à-vis the research questions and ecological systems at hand.

Under field conditions, the response of plants to changes in precipitation amount, seasonality, or frequency will depend on the physical and chemical characteristics of the soil, the response of the dominant and subdominant individual plants in the system, the current environmental conditions, and the characteristics of the potential future environment. One obvious thread through the chapters of this volume is the importance of timing, or *seasonality,* of precipitation relative to the actual *amount* of precipitation. Although the importance of precipitation seasonality in arid and semiarid ecosystems has been recognized for some time (e.g., Walter 1954, 1979), this volume provides strong empirical evidence of its importance to the structure and function of ecosystems as diverse as sagebrush steppe (see chapter 6), midgrass and tallgrass prairies (see chapters 7 and 9), oak savannas (see chapter 8), and eastern deciduous forest (see chapter 10). Relationships between precipitation seasonality (and the corresponding spatial and temporal distribution of soil moisture) and phenological and morphological adaptations of dominant plant functional types are reflected in regional-scale patterns of vegetation, from the grass-dominated Prairie Peninsula in the midwestern United States to lower-elevation ecotones of woody vegetation in the southwestern and interior northwestern United States (see chapter 4).

Linking Pattern to Process:
Experiments as Tools to Improve Ecological Understanding

Although this book does not describe all research that has been completed on manipulation of precipitation in North America (e.g., Reynolds et al. 1999) or on other continents (e.g., review by Gunderson et al. 1998; Grime et al. 2000), it should be clear that many important terrestrial systems remain relatively unstudied. The few systems that have been studied merit further investigation, but research is notably absent in coniferous woodlands and forests, shrublands, and warm deserts. Precipitation undoubtedly plays a major role in community- and ecosystem-level processes in many of these unexplored systems. The paucity of data from these and other systems constrains our ability to generalize about the response of species, growth-forms, life-forms, community-level properties (e.g., productivity, diversity), or ecosystem attributes (e.g., nutrient cycling, energy flows) to changing precipitation regimes. Our inability to generalize imposes restrictions on ecological understanding and effective management. As such, most management decisions and activities are dependent upon site-specific, descriptive research, which is only poorly suited to determine causal relationships (McPherson 1997; McPherson and Weltzin 2000). Thus, we argue that additional experimentally based research, comparable to that utilized in chapters 6 through 10, should be replicated and initiated in current and new terrestrial ecological systems, respectively.

In part because of the paucity of experimental, mechanistic research, the role of potential global and regional changes in precipitation regimes on plant interactions and population-, community-, and ecosystem-level processes remains largely unknown. Predictions of the distribution and composition of plant communities in response to potential changes in amount, frequency, and seasonality of precipitation are further hampered by the background of recurrent disturbances (such as harvesting, grazing, and fire) and the possible complexity of and general dearth of knowledge about regionally specific climate change (Houghton et al. 2001). However, the determination of most likely scenarios of climate change is relatively straightforward, and the effects of these changes on populations, communities, and ecosystems may be tested experimentally. Experiments that focus on interactions between various precipitation regimes and other climatic (temperature) and atmospheric (concentrations of greenhouse gases) factors are greatly needed, as they most likely will provide the greatest simultaneous contribution to ecological understanding and management.

Experiments on precipitation regimes should be designed to maximize their contribution to both the scientific and management communities. In particular, experiments should be designed to provide empirical data on soil-plant-water relationships, which form the basis for a mechanistic determination of plant recruitment, migration, and mortality of different plant functional types under changing climates (Walker 1996). Many such experiments produce unexpected responses of population, communities, and ecosystems to changing environmental conditions (e.g., Harte and Shaw 1995). In addition, the spatial and temporal scale of experiments should incorporate realistic "background" conditions such as herbivory, spatial variability in soils, and periodic drought. Designs should strive to link atmospheric conditions with precipitation regimes (e.g., by applying water when clouds are present and humidity is high) or at least to consider these constraints. Where possible, the effects of precipitation amount should be separated from effects of precipitation seasonality. Contemporary applications of stable-isotope techniques can be used to trace water use and assess integrated plant response to water stress.

Experiments also should be designed to facilitate the integration of evolutionary ecology, population ecology, ecosystem ecology, and resource management. Rather than tackling this daunting challenge simultaneously, initial efforts probably should focus on paired components (e.g., population-community, community-ecosystem). For example, determining the existence and relative importance of positive and negative feedback between plants, nutrients, energy, and carbon pools and fluxes may represent the greatest challenge to the integration of community and ecosystem ecology.

Explicitly incorporating linkages between population, community, and ecosystem patterns and processes may be prudent, due to the high economic cost of implementing and maintaining field experiments that manipulate precipitation. Thus, for example, community ecologists should seek collaborators interested in microbial ecology, ecophysiology, molecular ecology, etc. With only minor retooling, most community-level experimental research could also assess ecosystem-level response variables such as gross and net primary production, net ecosystem exchange of carbon, and pools and fluxes of major nutrients. With only minimal extra data collection, scientists could model soil water budgets, energy budgets, and plant demographics.

Further, the transfer of technology and ecological understanding to landscape managers, with due consideration for the juxtaposition between typical ecological and managerial spatial and temporal scales, will be critical. Managers of land and natural resources play a crucial role in the conservation of natural ecosystems, because they are accountable and responsible to a far greater extent than scientists. Management decisions often have far-reaching and lasting consequences. Thus, scientists have an obligation to conduct and synthesize research that is relevant to management and to make the results accessible to managers. A substantial infrastructure in the United States is well suited to facilitate the transfer of ecological knowledge: Cooperative Extension personnel are associated with each land-grant university and every county in the country, and the USDA Natural Resources Conservation Service provides assistance to private landowners.

To provide information to policymakers and the general public, research should consider the impacts of precipitation regime changes on the goods and services provided by ecosystems. For example, terrestrial ecosystems not only have intrinsic value, but also produce marketable products, are important for recreation and maintenance of species, contribute to aesthetic and spiritual experiences, and provide services such as the accumulation and cycling of nutrients, assimilation of atmospheric carbon dioxide and release of oxygen through respiration, sequestration of carbon, regulation of regional climates and hydrologic cycles, filtration of pollutants from air and water, and control of pests and pathogens (Costanza et al. 1997). Each of these uses and processes could be affected by changes in the ecosystem in response to variations in precipitation.

Future Research Directions

Interactive effects of soil moisture and temperature on interactions between plants, particularly of different growth-forms or life-forms, are largely unknown. For example, high temperatures of shallow soil layers may constrain root activity during the summer months and thereby limit uptake of water by tree roots to deeper soil depths, regardless of availability (Williams and Ehleringer 2000). This process could be governed by the amount and seasonality of precipitation and soil depth and texture. Hypotheses related to interactive effects of temperature and soil moisture on root initiation, distribution, and function can be readily investigated with field experiments similar to those described in this volume.

This volume focuses on the response of only one trophic level—primary producers—to changes in precipitation regimes. However, changes in precipitation will have direct effects on both consumers and decomposers, which undoubtedly will affect vegetation indirectly, though changes in rates of granivory, herbivory, nutrient cycling, and substrate alteration.

Predictions of future precipitation regimes depend on output from general circulation models (GCMs), which are being constantly improved. Most GCMs are parameterized at the global scale, with grid cells that can encompass entire regions, although a few have been executed at regional scales (e.g., Giorgi et al. 1998). Prediction of effects of precipitation change on vegetation will require additional development of local or regional models with output at a temporal resolution of months, or for temperate humid forests, at a temporal resolution of days (P. Hanson, pers. comm., 2001). Such scenarios could form the basis for new field experiments in ecosystems predicted to be highly sensitive to precipitation change (i.e., characterized by significant growth reductions or mortality). Where feasible and cost-effective, new experiments on changing precipitation regimes should include either elevated CO_2 or increased temperature, or both, to reflect the multiple interacting environmental changes that will coincide with global change. For example, ongoing research at Jasper Ridge, California, is investigating interactive effects of CO_2, temperature, water, and soil nitrogen on California annual grasslands (C. B. Field, pers. comm., 2001).

The relationship between GCMs and experiments should be reciprocal. GCM predictions can serve as most likely scenarios of climate change that delimit field experiments. In turn, results from field experiments should facilitate model parameterization. Additionally, GCMs should be linked with dynamic global vegetation models (DGVMs) to model feedback between terrestrial vegetation and climate (Neilson and Drapek 1998; Bachelet et al. 2001; Cramer et al. 2001; Neilson, chapter 4, this volume). Constructive interactions between modelers and empiricists will strengthen linkages between models and experiments, to the benefit of ecology and management. For example, model output from DGVMs that ignores edaphic factors represents little more than a computational return to Clements' (1916) concept of the climatic vegetation climax (see McAuliffe, chapter 2, this volume). In contrast, DGVMs that incorporate soils data reveal richer and more realistic relationships between potential future climates and vegetation (Bachelet et al. 1998).

Critical observations of alterations in species composition after environmental perturbations (e.g., Falkengren-Grerup and Eriksson 1990; Allen and Breshears 1998) will complement improved models of vegetation dynamics and enhance confidence predictions about the fate of communities and eco-systems over decadal temporal periods. Other topics that warrant additional investigation include the following: (1) role of precipitation and drought in controlling rates and patterns of vegetation change (e.g., Allen and Breshears 1998; Hanson and Weltzin 2000), (2) importance of soil water availability on structural and functional response of plant root allocation and architecture (e.g., Snyder and Williams 2000), (3) role of altered precipitation and episodic precipitation regimes on susceptibility of natural ecosystems to invasion by non-native plant species (Smith et al. 2000), (4) role and importance of pre-cipitation seasonality, serial correlation, and extremes on ecosystem structure and function (e.g., Easterling et al. 2000), and (5) interactive effects of changes in precipitation regimes with other anthropogenic and natural ecosystem stressors, including acid deposition, nutrient deposition, increases in tropo-spheric ozone, insects and other pests, and pathogens.

References

Alder, N. N., J. S. Sperry, and W. T. Pockman. 1996. Root and stem xylem embolism, stomatal conductance, and leaf turgor in *Acer grandidentatum* populations along a soil moisture gradient. *Oecologia* 105:293–301.

Allen, C. D., and D. D. Breshears. 1998. Drought-induced shift of a forest-woodland ecotone: Rapid landscape response to climate variation. *Proceedings of the National Academy of Science* 95:14839–14842.

Allison, G. B., C. J. Barnes, and M. W. Hughes. 1983. The distribution of deuterium and ^{18}O in dry soils. 2. Experimental. *Journal of Hydrology* 64:377–379.

Allmaras, R. R., and S. D. Logsdon. 1990. Soil structural influences on the root zone and rhizosphere. Pp. 8–54 *in* J. E. Box and L. C. Hammond, eds. *Rhizosphere dynamics.* AAAS Symposium Series. Westview Press, Boulder, Colo.

Amthor, J. S. 1995. Terrestrial higher-plant response to increasing atmospheric [CO_2] in relation to the global carbon cycle. *Global Change Biology* 1:243–274.

Anderson, R. S. 1989. Development of the southwestern ponderosa pine forests: What do we really know? Pp. 15–22 *in* A. Tecle, W. Covington, and R. H. Hamre, eds. *Multiresource management of ponderosa pine forests.* USDA Forest Service, Flagstaff, Ariz.

Archer, S. 1990. Development and stability of grass/woody mosaics in a subtropical savanna parkland, Texas, U.S.A. *Journal of Biogeography* 17:453–462.

———. 1995. Tree-grass dynamics in a *Prosopis*-thornscrub savanna parkland: Reconstructing the past and predicting the future. *Écoscience* 2:83–99.

Austin, M. 1903. *The land of little rain.* Houghton Mifflin Company, Boston.

Austin, M., O. B. Williams, and L. Bellin. 1981. Grassland dynamics under sheep grazing in an Australian Mediterranean-type climate. *Vegetation* 47:201–211.

Axelrod, D. I. 1985. Rise of the grassland biome, central North America. *Botanical Review* 51:163–201.

Bachelet D., M. Brugnach, and R. P. Neilson. 1998. Sensitivity of a biogeography model to soil properties. *Ecological Modeling* 109:77–98.

Bachelet, D., H. W. Hunt, and J. K. Detling. 1989. A simulation model of intraseasonal carbon and nitrogen dynamics of blue grama swards as influenced by above- and belowground grazing. *Ecological Modeling* 44:231–252.

Bachelet, D., and R. P. Neilson. 2000. Biome redistribution under climate change.

Pp. 18–44 *in* L. A. Joyce and R. Birdsey, tech. eds. *The impact of climate change on America's forests: A technical document supporting the* USDA *Forest Service* RPA *assessment.* General Technical Report RMRS-GTR-59. USDA Forest Service, Rocky Mountain Research Station, Fort Collins, Colo.

Bachelet, D., R. P. Neilson, J. M. Lenihan, and R. J. Drapek. 2001. Climate change effects on vegetation distribution and carbon budget in the United States. *Ecosystems* 4:164–185.

Bahre, C. J. 1991. *A legacy of change.* University of Arizona Press, Tucson.

Bailey, R. G. 1996. *Ecosystem geography.* Springer-Verlag, New York.

Bailey, R. G., P. E. Avers, T. King, and W. H. McNab, eds. 1994. Ecoregions and subregions of the United States (1:7,500,000 map). USDA Forest Service, Washington, D.C.

Barnes, C. J., and G. B. Allison. 1983. The distribution of deuterium and ^{18}O in dry soils. *Journal of Hydrology* 60:141–156.

———. 1988. Tracing of water movement in the unsaturated zone using stable isotopes of hydrogen and oxygen. *Journal of Hydrology* 100:143–176.

Barros, M. C., M. J. M. Mendo, and F. C. R. Negrao. 1995. Surface water quality in Portugal during a drought period. *Science of the Total Environment* 171:69–76.

Baskin, J. M., and C. C. Baskin. 1981. Ecology of germination and flowering in the weedy, winter, annual grass *Bromus japonicus. Journal of Range Management* 34:369–372.

Bates, J., T. Svejcar, R. Angell, and R. Miller. 1999. Changes in plant community dynamics resulting from altered precipitation. *In* W. C. Krueger, ed. *Biodiversity project annual report.* Department of Rangeland Resources, Oregon State University, Corvallis.

Beare, M. H., C. L. Neely, D. C. Coleman, and W. L. Hargrove. 1990. A substrate-induced respiration (SIR) method for measurement of fungal and bacterial biomass on plant residues. *Soil Biology and Biochemistry* 22:585–594.

Biondini, M. E., and L. Manske. 1996. Grazing frequency and ecosystem processes in a northern mixed prairie. *Ecological Applications* 6:239–256.

Biondini, M. E., B. D. Patton, and P. E. Nyren. 1998. Grazing intensity and ecosystem processes in a northern mixed-grass prairie. *Ecological Applications* 8:469–479.

Birkeland, P. W. 1984. *Soils and geomorphology.* Oxford University Press, New York.

Bittman, S., E. Z. Jan, and G. M. Simpson. 1987. Hand-operated rainout shelter. *Agronomy Journal* 79:1084–1086.

Bloom, A. J., I. F. S. Chapin, and H. A. Mooney. 1985. Resource limitation in plants: An economic analogy. *Annual Review of Ecology and Systematics* 16:363–392.

Boer, G. J., G. M. Flato, and D. Ramsden. 1999. A transient climate change simulation with historical and projected greenhouse gas and aerosol forcing: Projected climate for the 21st century. *Climate Dynamics* 16:427–450.

Boer, G. J., G. M. Flato, M. C. Reader, and D. Ramsden. 1999. A transient climate change simulation with historical and projected greenhouse gas and aerosol forcing: Experimental design and comparison with the instrumental record for the 20th century. *Climate Dynamics* 16:405–425.

Borchert, J. R. 1950. The climate of the central North American Grassland. *Annals of the Association of American Geographers* 40:1–39.

Breshears, D. D., and F. J. Barnes. 1999. Interrelationships between plant functional types and soil moisture heterogeneity for semiarid landscapes within the grassland/forest continuum: A unified conceptual model. *Landscape Ecology* 14:465–478.

Breshears, D. D., O. B. Myers, S. R. Johnson, C. W. Meyer, and S. N. Martens. 1997. Differential use of spatially heterogeneous soil moisture by two semiarid woody species: *Pinus edulis* and *Juniperus monosperma*. *Journal of Ecology* 85:289–299.

Breshears, D. D., J. W. Nyhan, C. E. Heil, and B. P. Wilcox. 1998. Effects of woody plants on microclimate in a semiarid woodland: Soil temperature and evaporation in canopy and intercanopy patches. *International Journal of Plant Science* 159:1010–1017.

Breshears, D. D., P. M. Rich, F. J. Barnes, and K. Campbell. 1997. Overstory-imposed heterogeneity in solar radiation and soil moisture in a semi-arid woodland. *Ecological Applications* 7:1201–1215.

Briggs, J. M., and A. K. Knapp. 1995. Interannual variability in primary production in tallgrass prairie: Climate, soil moisture, topographic position, and fire as determinants of aboveground biomass. *American Journal of Botany* 82:1024–1030.

Brown, D. E., ed. 1982. Biotic communities of the American southwest—United States and Mexico. *Desert Plants* 4:1–342.

Brown, J. H., T. J. Valone, and C. G. Curtin. 1997. Reorganization of an arid ecosystem in response to recent climate change. *Proceedings of the National Academy of Sciences of the United States of America* 94:9729–9733.

Brown, J. R., and S. Archer. 1989. Woody plant invasion of grasslands: establishment of honey mesquite (*Prosopis glandulosa* var. *glandulosa*) on sites differing in herbaceous biomass and grazing history. *Oecologia* 80:19–26.

——. 1990. Water relations of a perennial grass and seedling vs adult woody plants in a subtropical savanna, Texas. *Oikos* 57:366–374.

——. 1999. Shrub invasion of grassland: Recruitment is continuous and not regulated by herbaceous biomass or density. *Ecology* 80:2385–2396.

Bruce, R. R., and F. L. Shuman, Jr. 1962. Design for automatic movable plot shelters. *Transactions of the American Society of Agricultural Engineers* 52:212–213.

Brunel, J. P., G. R. Walker, and A. K. Kennett-Smith. 1995. Field validation of isotopic procedures for determining sources of water used by plants in a semi-arid environment. *Journal of Hydrology* 167:351–368.

Bryson, R. A. 1966. Air masses, streamlines, and the boreal forest. *Geographical Bulletin* 8:228–269.

Bryson, R. A., and F. K. Hare. 1974. Climates of North America. Pp. 1–47 *in* R. A. Bryson and F. K. Hare, eds. *World survey of climatology,* vol. 11. Elsevier Scientific Publishing Company, New York.

Buffington, L. C., and C. H. Herbel. 1965. Vegetational changes on a semidesert grassland range from 1858 to 1963. *Ecological Monographs* 35:139–164.

Bull, W. B. 1991. *Geomorphic responses to climatic change.* Oxford University Press, New York.

Burgess, S. O., M. A. Adams, N. C. Turner, and C. K. Ong. 1998. The redistribution of soil water by tree root systems. *Oecologia* 115:306–311.

Burgess, S. S. O., M. A. Adams, N. C. Turner, D. A. White, and C. K. Ong. 2001. Tree roots: Conduits for deep recharge of soil water. *Oecologia* 126:158–165.

Burgess, T. L. 1995. Desert grassland, mixed shrub savanna, shrub steppe, or semidesert scrub? The dilemma of coexisting growth forms. Pp. 31–67 *in* M. P. McClaran and T. R. Van Devender, eds. *The desert grassland.* University of Arizona Press, Tucson.

Busch, D. E., N. L. Ingraham, and S. D. Smith. 1992. Water uptake in woody riparian phreatophytes of the southwestern United States: A stable isotope study. *Ecological Applications* 2:450–459.

Busso, C. A., and J. H. Richards. 1993. Leaf extension rate in two tussock grasses: Effects of growth of two tussock grasses in Utah. *Journal of Arid Environments* 29:239–251.

Butler, J. L., and D. D. Briske. 1988. Population structure and tiller demography of the bunchgrass *Schizachyrium scoparium* in response to herbivory. *Oikos* 51:306–312.

Cable, D. R. 1975. Influence of precipitation on perennial grass production in the semidesert southwest. *Ecology* 56:981–986.

Caldwell, M. M. 1979. Physiology of sagebrush. Pp. 74–85 *in Sagebrush ecosystem: A symposium.* Utah State University, College of Natural Resources, Logan.

Caldwell, M. M., T. E. Dawson, and J. H. Richards. 1998. Hydraulic lift: Consequences of water efflux from the roots of plants. *Oecologia* 113:151–161.

Caprio, A. C., and M. J. Zwolinski. 1995. Fire and vegetation in a madrean oak woodland, Santa Catalina Mountains, southeastern Arizona. Pp. 389–399 *in* L. F. DeBano, G. J. Gottfried, R. H. Hamre, C. B. Edminster, P. F. Ffolliott, and A. Ortega-Rubio, tech. coordinators. *Biodiversity and management of the*

Madrean Archipelago: The sky islands of southwestern United States and Northwestern Mexico. USDA Forest Service General Technical Report RM-264. Rocky Mountain Forest and Range Experiment Station, Fort Collins, Colo.

Carlson, D. H., T. L. Thurow, R. W. Knight, and R. K. Heitschmidt. 1990. Effect of honey mesquite on the water balance of Texas Rolling Plains rangeland. *Journal of Range Management* 43:491–496.

Carmichael, R. S., O. D. Knipe, C. P. Pase, and W. W. Brady. 1978. Arizona chaparral: Plant associations and ecology. P. 16 *in* USDA Forest Service Research Paper RM-202. Rocky Mountain Forest and Range Experiment Station, Fort Collins, Colo.

Ceulemans, R. J. M., A. Janssens, and M. E. Jach. 1999. Effects of CO_2 enrichment on trees and forests: Lessons to be learned in view of future ecosystem studies. *Annals of Botany* 84:577–590.

Chesson, P., and N. Huntley. 1989. Short-term instabilities and long-term community dynamics. *Trends in Ecology and Evolution* 4:293–298.

Clark, G. A., and D. L. Reddell. 1990. Construction details and microclimate modifications of a permanent rain-sheltered lysimeter system. *Transactions of the American Society of Agricultural Engineers* 33:1813–1822.

Clary, W. P, and A. R. Tiedemann. 1986. Distribution of biomass in small tree and shrub form *Quercus gambelii* stands. *Forest Science* 32:234–242.

Clawson, K. L., B. L. Blad, and J. E. Specht. 1986. Use of portable rainout shelters to induce water stress. *Agronomy Journal* 78:120–123.

Clements, F. E. 1916. *Plant succession.* Carnegie Institute of Washington Publication 242. Carnegie Institute of Washington, Washington, D.C.

Collins, S. L., and S. M. Glenn. 1991. Importance of spatial and temporal dynamics in species regional abundance and distribution. *Ecology* 72: 654–664.

Cook, E. R., M. A. Kablack, and G. C. Jacoby. 1988. The 1986 drought in the southeastern United States: How rare an event was it? *Journal of Geophysical Research* 93(D11):14257–14260.

Cook, J. G., and L. L. Irwin. 1992. Climate-vegetation relationships between the Great Plains and Great Basin. *American Midland Naturalist* 127:316–326.

Costanza, R., R. d'Arge, R. de Groot, S. Farber, M. Grasso, B. Hannon, K. Limburg, S. Naeem, R. V. O'Neill, J. Paruelo, R. G. Raskin, P. Sutton, and M. van den Belt. 1997. The value of the world's ecosystem services and natural capital. *Nature* 387:253–260.

Cottam, W. P., J. M. Tucker, and R. Drobnick. 1959. Some clues to Great Basin postpluvial climates provided by oak distributions. *Ecology* 40:361–377.

Coughenour, M. B., and W. J. Parton. 1996. Integrated models of ecosystem function: A grassland case study. Pp. 93–114 *in* B. H. Walker and W. L. Steffen, eds. *Global change and terrestrial ecosystems.* Cambridge University Press, Cambridge.

Coupland, R. T. 1958. The effects of fluctuations in weather upon the grasslands of the Great Plains. *Botanical Review* 24:273–317.

———. 1979. The nature of grassland. Pp. 23–29 *in* R. T. Coupland, ed. *Grassland ecosystems of the world: Analysis of grasslands and their uses*. Cambridge University Press, Cambridge.

Cramer, W., A. Bondeau, F. I. Woodward, I. C. Prentice, R. A. Betts, V. Brovkin, P. M. Cox, V. Fisher, J. A. Foley, A. D. Friend, C. Kucharik, M. R. Lomas, N. Ramankutty, S. Sitch, B. Smith, A. White, and C. Young-Molling. 2001. Global response of terrestrial ecosystem structure and function to CO_2 and climate change: Results from six dynamic global vegetation models. *Global Change Biology* 7:357–373.

Cronquist, A., A. H. Holmgren, N. H. Holmgren, J. L. Reveal, and P. K. Holmgren. 1977. *Intermountain flora*. New York Botanical Garden, Bronx.

Dadkhah, M., and G. F. Gifford. 1980. Influence of vegetation, rock cover, and trampling on infiltration rates and sediment production. *Water Resource Bulletin* 16:979–986.

Daly, C., D. Bachelet, J. M. Lenihan, W. Parton, R. P. Neilson, and D. Ojima. 2000. Dynamic simulation of tree-grass interactions for global change studies. *Ecological Applications* 10:449–469.

Dansgaard, W. 1964. Stable isotopes in precipitation. *Tellus* 16:436–468.

Davis, M. B. 1989. Lags in vegetation response to greenhouse warming. *Climatic Change* 15:75–82.

Davis, O. K. 1994. The correlation of summer precipitation in the southwestern U.S.A. with isotopic records of solar activity during the Medieval Warm Period. *Climatic Change* 26:271–287.

Dawson, T. E. 1993. Hydraulic lift and water use by plants: Implications for water balance, performance and plant-plant interactions. *Oecologia* 95:565–574.

Dawson, T. E., and J. R. Ehleringer. 1991. Streamside trees that do not use stream water. *Nature* 350:335–337.

———. 1993. Isotopic enrichment of water in the woody tissues of plants: Implications for plant water source, water uptake, and other studies which use stable isotopes. *Geochemica et Cosmochemica Acta* 57:3487–3492.

Diamond, D. D., and F. E. Smeins. 1988. Gradient analysis of remnant true and upper coastal prairie grasslands of North America. *Canadian Journal of Botany* 66:2152–2161.

Díaz, S. 1995. Elevated CO_2 responsiveness, interactions at the community level and plant functional types. *Journal of Biogeography* 22:289–295.

Dodd, J. L., W. K. Lauenroth, and R. K. Heitschmidt. 1982. Effects of controlled SO_2 exposure on net primary production and plant biomass dynamics. *Journal of Range Management* 35:572–579.

Donovan, L. A., and J. R. Ehleringer. 1992. Contrasting water-use patterns among size and life-history classes of a semi-arid shrub. *Functional Ecology* 6:482–488.

Drungil, C. E. C., T. J. Gish, and K. Abt. 1987. *Soil moisture determination in gravelly soils with time domain reflectometry.* Paper Number 87–2568. American Society of Agricultural Engineers, St. Joseph, Mich.

Dugas, W. A., Jr., and D. R. Upchurch. 1984. Microclimate of a rainfall shelter. *Agronomy Journal* 76:867–871.

Dukes, J. S., and H. A. Mooney. 1999. Does global change increase the success of biological invaders? *Trends in Ecology and Evolution* 14:135–139.

Easterling, W. E. 1990. Climate trends and prospects. Pp. 32–55 *in* R. N. Sampson and D. Hair, eds. *Natural resources for the 21st century.* Island Press, Washington, D.C.

Easterling, W. E., G. A. Meehl, C. Parmesan, S. A. Changnon, T. R. Karl, and L. O. Mearns. 2000. Climate extremes: Observations, modeling, and impacts. *Science* 289:2068–2074.

Eberhardt, L. L., and J. M. Thomas. 1991. Designing environmental field studies. *Ecological Monographs* 61:53–73.

Ehleringer, J. R., and T. E. Dawson. 1992. Water uptake by plants: Perspectives from stable isotope composition. *Plant Cell and Environment* 15:1073–1082.

Ehleringer, J. R., and S. L. Phillips. 1996. Ecophysiological factors contributing to the distributions of several *Quercus* species in the intermountain west. *Annales des Sciences Forestières* 53:291–302.

Ehleringer, J. R., S. L. Phillips, W. S. F. Schuster, and D. R. Sandquist. 1991. Differential utilization of summer rains by desert plants. *Oecologia* 88:430–434.

Eissenstat, D. M., and R. D. Yanai. 1997. The ecology of root lifespan. *Advances in Ecological Research* 27:1–60.

Eldrige, D., and D. Freudenberger, eds. 1999. *People and rangelands building the future: Proceedings of the VI International Rangeland Congress, Townsville, Queensland, Australia.* Elect Printing, Fyshwick.

Emmanuel, W. R., H. H. Shugart, and M. Stevenson. 1985. Climatic change and the broad scale distribution of terrestrial ecosystem complexes. *Climatic Change* 7:29–43.

Emmerich, W., and R. K. Heitschmidt. 2002. Drought and grazing: II. Effects on quantity and quality of water. *Journal of Range Management,* in press.

Eneboe, E. J. 1996. Tiller dynamics of blue grama and western wheatgrass subjected to drought and grazing. Master's thesis, Montana State University, Bozeman.

Epstein, H. E., W. K. Lauenroth, and I. C. Burke. 1997. Effects of temperature and soil texture on ANPP in the U.S. Great Plains. *Ecology* 78:2628–2631.

Ernest, S. K. M., J. H. Brown, and R. R. Parmenter. 2000. Rodents, plants, and precipitation: Spatial and temporal dynamics of consumers and resources. *Oikos* 88:470–482.

Ettershank, G., J. Ettershank, M. Bryant, and W. G. Whitford. 1978. Effects of nitrogen fertilization on primary production in a Chihuahuan desert ecosystem. *Journal of Arid Environments* 1:135–139.

Evans, R. D., R. A. Black, and S. O. Link. 1991. Reproductive growth during drought in *Artemisia tridentata. Functional Ecology* 5:676–683.

Evans, R. D., and J. R. Ehleringer. 1994. Water and nitrogen dynamics in an arid woodland. *Oecologia* 99:233–242.

Falkengren-Grerup, U., and H. Eriksson. 1990. Changes in soil, vegetation, and forest yield between 1947 and 1988 in beech and oak sites of southern Sweden. *Forest Ecology and Management* 38:37–53.

Fay, P. A., J. D. Carlisle, A. K. Knapp, J. M. Blair, and S. L. Collins. 2000. Altering rainfall timing and quantity in a mesic grassland ecosystem: Design and performance of rainfall manipulation shelters. *Ecosystems* 3:308–319.

Field, C. B., R. B. Jackson, and H. A. Mooney. 1995. Stomatal responses to increased CO_2: Implications from the plant to the global scale. *Plant, Cell and Environment* 18:1214–1225.

Field, C. B., C. P. Lund, N. R. Chiariello, and B. E. Mortimer. 1997. CO_2 effects on the water budget of grassland microcosm communities. *Global Change Biology* 3:197–206.

Flanagan, L. B., J. R. Ehleringer, and J. D. Marshall. 1992. Differential uptake of summer precipitation among co-occurring trees and shrubs in a pinyon-juniper woodland. *Plant, Cell and Environment* 15:831–836.

Flato, G. M., G. J. Boer, W. G. Lee, N. A. McFarlane, D. Ramsden, M. C. Reader, and A. J. Weaver. 1999. The Canadian Centre for Climate Modelling and Analysis global coupled model and its climate. *Climate Dynamics* 16:451–467.

Flerchinger, G. N., and K. E. Saxton. 1989a. Simultaneous heat and water model of a freezing snow-residue-soil system. I. Theory and development. *Transactions of the American Society of Agricultural Engineers* 32:565–571.

——. 1989b. Simultaneous heat and water model of a freezing snow-residue-soil system. II. Field verification. *Transactions of the American Society of Agricultural Engineers* 32:573–578.

Foale, M. A., R. Davis, and D. R. Upchurch. 1986. The design of rain shelters for field experimentation: A review. *Journal of Agricultural Engineering Research* 34:1–16.

Foran, B. D., G. Bastin, E. Remenga, and K. W. Hyde. 1982. The response to season, exclosure and distance from water of three central Australian pasture types grazed by cattle. *Australian Rangeland Journal* 4:5–15.

Fredeen, A. L., J. T. Randerson, N. M. Holbrook, and C. B. Field. 1997. Elevated atmospheric CO_2 increases water availability in a water-limited grassland ecosystem. *Journal of the American Water Resources Association* 33:1033–1039.

Garten, C. T., Jr., M. A. Huston, and C. A. Thoms. 1994. Topographic variation of soil nitrogen dynamics at Walker Branch Watershed, Tennessee. *Forest Science* 40:497–512.

Gat, J. R. 1980. Groundwater. Pp. 223–240 *in* J. R. Gat and R. Gonfiantini, eds. *Stable isotope hydrology: Deuterium and oxygen-18 in the water cycle.* IAEA Technical Series Report Number 210. IAEA, Vienna.

Georgiadis, N. J., R. W. Ruess, S. J. McNaughton, and D. Western. 1989. Ecological conditions that determine when grazing stimulates grass production. *Oecologia* 81:316–322.

Germaine, H. L., and G. R. McPherson. 1999. Effects of biotic factors on emergence and survival of *Quercus emoryi* at lower treeline. *Écoscience* 6:92–99.

Gibson, A. C., and P. S. Nobel. 1986. *The cactus primer.* Harvard University Press, Cambridge.

Gile, L. H. 1975a. Causes of soil boundaries in an arid region: I. Age and parent materials. *Soil Science Society of America Proceedings* 39:316–323.

———. 1975b. Causes of soil boundaries in an arid region: II. Dissection, moisture, and faunal activity. *Soil Science Society of America Proceedings* 39:324–330.

Gile, L. H., J. W. Hawley, and R. B. Grossman. 1981. *Soils and geomorphology in the basin and range area of Southern New Mexico—Guidebook to the desert project.* New Mexico Bureau of Mines and Mineral Resources, Socorro.

Giorgi, F., C. S. Brodeur, and G. T. Bates. 1994. Regional climate change scenarios over the United States produced with a nested vegetation climate model. *Journal of Climate* 7:375–399.

Giorgi, F., L. O. Mearns, C. Shields, and L. McDaniel. 1998. Regional nested model simulations of present day and $2xCO_2$ climate over the central plains of the United States. *Climatic Change* 40:457–493.

Glendening, G. E., and H. A. Paulsen. 1955. *Reproduction and establishment of velvet mesquite as related to invasion of semidesert grasslands.* USDA Technical Bulletin 1127. U.S. Government Printing Office, Washington, D.C.

Goldberg, D. E., and T. E. Miller. 1990. Effects of different resource additions on species diversity in an annual plant community. *Ecology* 71:213–225.

Goldstein, G. H., and G. Sarmiento. 1991. Water relations of trees and grasses and their consequences for the structure of savanna vegetation. Pp. 13–38 *in* B. Walker, ed. *Determinants of tropical savannas.* IUBS Monograph Series. IUBS, Oxford, England.

Graham, R. W., and E. C. Grimm. 1990. Effects of global climate change on the patterns of terrestrial biological communities. *Trends in Ecology and Evolution* 5:289–292.

Gregory, J. M., J. F. B. Mitchell, and A. J. Brady. 1997. Summer drought in northern mid-latitudes in a time-dependent CO$_2$ climate experiment. *Journal of Climate* 10:662–686.

Gregory, P. J., J. S. I. Ingram, B. Campbell, J. Goudriaan, L. A. Hunt, J. J. Landsberg, S. Linder, M. Stafford-Smith, R. W. Sutherst, and C. Valentin. 1999. Managed production systems. Pp. 229–270 *in* B. Walker, W. Steffen, J. Canadell, and J. Ingram, eds. *The terrestrial biosphere and global change: Implications for natural and managed systems.* Cambridge University Press, Cambridge.

Grime, J. P. 1979. *Plant strategies and vegetation processes.* John Wiley and Sons, Chichester.

Grime, J. P., V. K. Brown, K. Thompson, G. J. Masters, S. H. Hiller, I. P. Clarke, A. P. Askew, D. Corker, and J. P. Kielty. 2000. The response of two contrasting limestone grasslands to simulated climate change. *Science* 289:762–765.

Grover H. D., and H. B. Musick. 1990. Shrubland encroachment in southern New Mexico, U.S.A.: An analysis of desertification processes in the American Southwest. *Climate Change* 17:305–330.

Grubb, P. J. 1977. The maintenance of species richness in plant communities: The importance of the regeneration niche. *Biological Review* 52:107–145.

Gunderson, P., A. W. Boxman, N. Lamersdorf, F. Moldan, and B. R. Andersen. 1998. Experimental manipulation of forest ecosystems: Lessons from large roof experiments. *Forest Ecology and Management* 101:339–352.

Hacke, U. G., J. S. Sperry, B. E. Ewers, D. S. Ellsworth, K. V. R. Schäfer, and R. Oren. 2000. Influence of soil porosity on water use in *Pinus taeda. Oecologia* 124:495–505.

Haferkamp, M. R., P. O. Currie, J. D. Volesky, and B. W. Knapp. 1992. Mortality of crested wheatgrass and Russian wildrye during drought. *Journal of Range Management* 45:355–357.

Haferkamp, M. R., M. G. Karl, and M. D. MacNeil. 1994. Influence of storage, temperature, and light on germination of Japanese brome seed. *Journal of Range Management* 47:140–144.

Haferkamp, M. R., M. D. MacNeil, and M. G. Karl. 1996. Induction of secondary dormancy in Japanese brome *(Bromus japonicus).* Pp. 199–200 *in* N. West, ed. *Proceedings of the Vth International Rangeland Congress, July 23–28, 1995, Salt Lake City.* Society for Range Management, Denver, Colo.

Haferkamp, M. R., D. Palmquist, J. A. Young, and M. D. MacNeil. 1995. Influence of temperature on germination of Japanese brome seed. *Journal of Range Management* 48:264–266.

Hake, D. R., J. Powell, J. K. McPherson, P. L. Claypool, and G. L. Dunn. 1984. Water stress of tallgrass prairie plants in central Oklahoma. *Journal of Range Management* 37:147–151.

Hall, D. O., and J. M. O. Scurlock. 1991. Climate change and productivity of natural grasslands. *Annals of Botany* 7:49–55.

Hanson, J. D., B. B. Baker, and R. M. Bourdan. 1993. Comparison of the effects of different climate change scenarios on rangeland livestock production. *Agricultural Systems* 41:487–502.

Hanson, P. J. 2000. Large-scale water manipulations. Pp. 341–352 *in* O. E. Sala, R. B. Jackson, H. A. Mooney, and R. W. Howarth, eds. *Methods in ecosystem science.* Springer, New York.

Hanson, P. J., D. E. Todd, and J. S. Amthor. 2001. A six year study of sapling and large-tree growth and mortality responses to natural and induced variability in precipitation and throughfall. *Tree Physiology* 21:345–358.

Hanson, P. J., D. E. Todd, N. T. Edwards, and M. A. Huston. 1995. Field performance of the Walker Branch Throughfall Displacement Experiment. Pp. 307–313 *in* A. Jenkins, R. C. Ferrier, and C. Kirby, eds. *Ecosystem manipulation experiments: Scientific approaches, experimental design, and relevant results.* Ecosystem Research Report Number 20. Commission of the European Communities, Brussels, Belgium.

Hanson, P. J., D. E. Todd, M. A. Huston, J. D. Joslin, J. Croker, and R. M. Augé. 1998. *Description and field performance of the Walker Branch Throughfall Displacement Experiment: 1993–1996.* ORNL Technical Memorandum 13586. Oak Ridge National Laboratory, Oak Ridge, Tenn.

Hanson, P. J., and J. F. Weltzin. 2000. Drought disturbance from climate change: Response of United States forests. *Science of the Total Environment* 262:205–220.

Harper, J. L. 1977. *Population biology of plants.* Academic Press, London.

Harrington, G. N. 1991. Effects of soil moisture on shrub seedling survival in a semi-arid grassland. *Ecology* 72:1138–1149.

Harrington, M. G., and R. G. Kelsey. 1979. *Influence of some environmental factors on initial establishment and growth of ponderosa pine seedlings.* USDA Forest Service Research Paper INT-230:1–26. Intermountain Range and Forest Experiment Station, Ogden, UT.

Harte, J., and R. Shaw. 1995. Shifting dominance within a montane vegetation community: Results of a climate-warming experiment. *Science* 267:876–880.

Harte, J., M. S. Torn, F. R. Chang, B. Feifarek, A. P. Kinzig, R. Shaw, and K. Shen. 1995. Global warming and soil microclimate: Results from a meadow-warming experiment. *Ecological Applications* 5:132–150.

Hartnett, D. C., and P. A. Fay. 1998. Plant populations: Patterns and processes. Pp. 81–100 *in* A. K. Knapp, J. M. Briggs, D. C. Hartnett, S. L. Collins, eds. *Grassland dynamics: Long-term ecological research in tallgrass prairie.* Oxford University Press, New York.

Hatfield, P. M., G. C. Wright, and W. R. Tapsall. 1990. A large, retractable, low cost, and re-locatable rain out shelter design. *Exploratory Agriculture* 26:57–62.

Haworth, K., and G. R. McPherson. 1994. Effects of *Quercus emoryi* on herbaceous vegetation in a semi-arid savanna. *Vegetatio* 112:153–159.

Hayden, B. P. 1998. Regional climate and the distribution of tallgrass prairie. Pp. 19–34 *in* A. K. Knapp, J. M. Briggs, D. C. Hartnett, and S. L. Collins, eds. *Grassland dynamics: Long-term ecological research in tallgrass prairie.* Oxford University Press, New York.

Hayes, D. C., and T. R. Seastedt. 1987. Root dynamics of tallgrass prairie in wet and dry years. *Canadian Journal of Botany* 65:787–791.

Heitschmidt, R. K., E. E. Grings, M. R. Haferkamp, and M. G. Karl. 1995. Herbage dynamics on two northern Great Plains range sites. *Journal of Range Management* 48:211–217.

Heitschmidt, R. K., M. R. Haferkamp, M. G. Karl, and A. L. Hild. 1999. Drought and grazing: I. Effects on quantity of forage produced. *Journal of Range Management* 52:440–446.

Heitschmidt, R. K., and J. W. Stuth, eds. 1991. Grazing management: An ecological perspective. Timber Press, Portland, Oreg.

Helms, S., R. Mendelsohn, J. Neumann. 1996. The impact of climate change on agriculture. *Climate Change* 33:1–6.

Hild, A. L., M. G. Karl, M. R. Haferkamp, and R. K. Heitschmidt. 2001. Drought and grazing: III. Root growth and germinable seed banks. *Journal of Range Management* 54:292–298.

Hiler, E. A. 1969. Quantitative evaluation of crop-drainage requirements. *Transactions of the American Society of Agricultural Engineers* 12:499–505.

Holdridge, L. R. 1947. Determination of world plant formations from simple climatic data. *Science* 105:367–368.

Holmgren, M. 1996. Interactive effect of shade and drought on seedling growth and survival. Ph.D. diss., University of Tennessee, Knoxville.

Horton, M. L. 1962. Rainout shelter for corn. *Iowa Farm Science* 17:16.

Houghton, J. T. 1997. *Global warming: The complete briefing.* Cambridge University Press, Cambridge

Houghton, J. T., Y. Ding, D. J. Griggs, M. Noguer, P. J. van der Linden, X. Dai, K. Maskell, and C. A. Johnson. 2001. *Climate change 2001: The scientific basis.* Contribution of Working Group 1 to the Third Assessment Report of the Intergovernmental Panel on Climate Change. Cambridge University Press, Cambridge.

Houghton, J. T., L. G. M. Filho, B. A. Callander, N. Harris, A. Kattenberg, and K. Maskell, eds. 1996. *Climate change 1995: The science of climate change.* Cambridge University Press, Cambridge.

Houghton, J. T., G. J. Jenkins, and J. J. Ephraums, eds. 1990. *Climate change. The* IPCC *Scientific Assessment.* Cambridge University Press, Cambridge.

Hsiao, T. C., W. K. Silk, and J. Jing. 1985. Leaf growth and water deficits: Biophysical effects. Pp. 239–666 *in* N. R. Baker, W. J. Davies, and C. K. Ong, eds. *Control of leaf growth.* Cambridge University Press, Cambridge.

Hubbard, J. A., and G. R. McPherson. 1999. Do seed predation and dispersal limit downslope movement of a semi-desert grassland/oak woodland transition? *Journal of Vegetation Science* 10:739–744.

Huck, M. G., and G. Hoogenboom. 1990. Soil and plant root water flux in the rhizosphere. Pp. 268–312 *in* J. E. Box and L. C. Hammond, eds. *Rhizosphere dynamics.* American Association for the Advancement of Science Symposium Series. Westview Press, Boulder, Colo.

Hultine, K., D. G. Williams, and S. S. O. Burgess. n.d. Hydraulic redistribution and diurnal storage capacitance in desert phreatophytes. *Oecologia,* submitted.

Hunt, E. R., Jr., N. J. D. Zakir, and P. S. Nobel. 1987. Water costs and water revenues for established and rain-induced roots of *Agave deserti. Functional Ecology* 1:125–129.

Hurlbert, S. H. 1984. Pseudoreplication and the design of ecological field experiments. *Ecological Monographs* 54:187–211.

Huston, M. A. 1994. Biological diversity: The coexistence of species on changing landscapes. Cambridge University Press, Cambridge.

Ingraham, N. L., B. F. Lyles, R. L. Jacobson, and J. W. Hess. 1991. Stable isotopic study of precipitation and spring discharge in southern Nevada. *Journal of Hydrology* 125:243–258.

Ingraham, N. L., and B. E. Taylor. 1991. Light stable isotope systematics of large-scale hydrologic regimes in California and Nevada. *Water Resources Research* 27:77–90.

Iverson, L. R., and A. M. Prasad. 1998. Predicting abundance of 80 tree species following climate change in the eastern United States. *Ecological Monographs* 68:465–485.

Jackson, R. B., J. Canadell, J. R. Ehleringer, H. A. Mooney, O. E. Sala, and E. D. Schulze. 1996. A global review of rooting patterns: I. Root distribution by depth for terrestrial biomes. *Oecologia* 108:389–411.

Jackson, R. B., H. J. Schenk, E. G. Jobbágy, J. Canadell, G. D. Colello, R. E. Dickenson, C. B. Field, P. Friedlingstein, M. Heimann, K. Hibbard, D. W. Kicklighter, A. Kleidon, R. P. Neilson, W. J. Parton, O. E. Sala, and M. T. Sykes. 2000. Belowground consequences of vegetation change and their treatment in models. *Ecological Applications* 10:470–483.

Jacoby, P. W., R. J. Ansley, and B. K. Lawrence. 1988. Design of rain shelters for studying water relations of rangeland shrubs. *Journal of Range Management* 41:83–85.

Jennings, C. M. H. 1974. The hydrology of Botswana. Ph.D. diss., University of Natal, South Africa.

Jesko, T., J. Navara, and K. Dekankova. 1997. Root growth and water uptake by flowering maize plants, under drought conditions. Pp. 270–271 *in* A. Altman and Y. Waisel, eds. *Biology of root formation*. Proceedings of the 2nd International Symposium on Biology of Root Formation and Development, June 23–28, 1996, Jerusalem, Israel. Plenum Press, New York.

Joffre, R., and S. Rambal. 1988. Soil water improvement by trees in the rangelands of Southern Spain. *Oecologia Plantarum* 9:405–422.

Johns, T. C., R. E. Carnell, J. F. Crossley, J. M. Gregory, J. F. B. Mitchell, C. A. Senior, S. F. B. Tett, R. A. Wood. 1997. The second Hadley Centre coupled ocean-atmosphere GCM: Model description, spinup, and validation. *Climate Dynamics* 13:103–134.

Johnson, D. A., K. H. Asay, L. Tieszen, J. R. Ehleringer, and P. G. Jefferson. 1990. Carbon isotope discrimination: Potential in screening cool-season grasses for water-limited environments. *Crop Science* 30:338–343.

Johnson, D. A., and M. D. Rumbaugh. 1995. Genetic variation and inheritance characteristics for carbon isotope discrimination in alfalfa. *Journal of Range Management* 48:126–131.

Johnson, D. W., P. J. Hanson, and D. E. Todd. 2002. The effects of throughfall manipulation on soil solution chemistry and leaching in a deciduous forest. *Journal of Environmental Quality* 31:204–216.

Johnson, D. W., P. J. Hanson, D. E. Todd, R. B. Susfalk, and C. Trettin. 1998. Precipitation change and soil leaching: field results and simulations from Walker Branch Watershed, Tennessee. *Water, Air and Soil Pollution* 105:251–262.

Johnson, D. W., and S. E. Lindberg, eds. 1992. *Atmospheric deposition and nutrient cycling in forest ecosystems*. Springer-Verlag, New York.

Johnson, D. W., R. B. Susfalk, H. L. Gholz, and P. J. Hanson. 2000. Simulated effects of temperature and precipitation change on several forest ecosystems. *Journal of Hydrology* 235:183–204.

Johnson, D. W., and R. I. van Hook, eds. 1989. Analysis of biogeochemical cycling processes in Walker Branch Watershed. Springer-Verlag, New York.

Joslin, J. D., and M. H. Wolfe. 1998. Impacts of long–term water input manipulations on fine root production and mortality in mature hardwood forests. *Plant and Soil* 204:165–174.

Joslin, J. D., M. H. Wolfe, and P. J. Hanson. 2000. Effects of shifting water regimes on forest root systems. *New Phytologist* 147:117–129.

——. 2001. Factors controlling the timing of root elongation intensity in a mature upland oak stand. *Plant and Soil* 228:201–212.

Joyce, L. A., and R. Birdsey, tech. eds. 2000. *The impact of climate change on America's forests: A technical document supporting the USDA Forest Service RPA assessment.* General Technical Report RMRS-GTR-59. USDA Forest Service, Rocky Mountain Research Station, Fort Collins, Colo.

Karl, M. G., R. K. Heitschmidt, and M. R. Haferkamp. 1999. Vegetation biomass dynamics and patterns of sexual reproduction in a northern mixed-grass prairie. *American Midland Naturalist* 141:227–237.

Karl, T. R., R. W. Knight, D. R. Easterling, and R. G. Quayle. 1996. Indices of climate change for the United States. *Bulletin of the American Meteorological Society* 77:279–292.

Kattenberg, A., F. Giorgi, H. Grassl, G. A. Meehl, J. F. B. Mitchell, R. J. Stouffer, T. Tokioka, A. J. Weaver, and T. M. L. Wigley. 1996. Climate models: Projections of future climate. Pp. 285–357 *in* J. T. Houghton, G. Miera, B. Filho, B. A. Callander, N. Harris, A. Kattenberg, and K. Maskell, eds. *Climate change 1995: The science of climate change.* Cambridge University Press, Cambridge.

Kemp, P. R. 1983. Phenological patterns of Chihuahuan desert plants in relation to the timing of water availability. *Journal of Ecology* 71:427–436.

Kemp, P. R., J. F. Reynolds, Y. Pachepsky, and J. Chen. 1997. A comparative modeling study of soil water dynamics in a desert ecosystem. *Water Resources Research* 33:73–90.

Kinucan, R. J., and F. E. Smeins. 1992. Soil seed bank of a semiarid Texas grassland under three long-term (36 years) grazing regimes. *American Midland Naturalist* 128:11–21.

Kirschbaum, M. U. F., and A. Fischlin. 1996. Climate change impacts on forests. Pp. 95–129 *in* R. T. Watson, M. C. Zinyowera, and R. H. Moss, eds. *Climate change 1995 impacts, adaptations and mitigation of climate change: Scientific-technical analysis.* Cambridge University Press, New York.

Kittel, T. G. F., N. A. Rosenbloom, T. H. Painter, D. S. Schimel, and VEMAP Modeling Participants. 1995. The VEMAP integrated database for modeling United States ecosystem/vegetation sensitivity to climate change. *Journal of Biogeography* 22:857–862.

Knapp, A. K. 1984. Water relations and growth of three grasses during wet and drought years in a tallgrass prairie. *Oecologia* 65:35–43.

Knapp, A. K., J. M. Briggs, D. C. Hartnett, and S. L. Collins. 1998a. *Grassland dynamics: Long-term ecological research in tallgrass prairie.* Oxford University Press, New York.

Knapp, A. K., S. L. Conard, and J. M. Blair. 1998. Determinants of soil CO_2 flux from a sub-humid grassland: Effect of fire and fire history. *Ecological Applications* 8:760–770.

Knapp, A. K., and J. T. Fahnestock. 1990. Influence of plant size on the carbon and water relations of *Cucurbita foetidissima* HBK. *Functional Ecology* 4:789–797.

Knapp, A. K., J. T. Fahnestock, S. P. Hamburg, L. B. Statland, T. R. Seastedt, and D. S. Schimel. 1993. Landscape patterns in soil-plant water relations and primary production in tallgrass prairie. *Ecology* 74:549–560.

Knapp, A. K., and E. Medina. 1999. Success of C_4 photosynthesis in the field: Lessons from communities dominated by C_4 plants. Pp. 251–283 *in* R. F. Sage and R. K. Monson, eds. *C_4 plant biology*. Academic Press, New York.

Knoop, W. T., and B. H. Walker. 1985. Interactions of woody and herbaceous vegetation in a southern African savanna. *Journal of Ecology* 73:235–253.

Koch, G. W., and H. A. Mooney. 1996. *Carbon dioxide and terrestrial ecosystems*. Academic Press, San Diego.

Kolb, T. E., S. C. Hart, and R. Amundson. 1997. Boxelder water source and physiology at perennial and ephemeral stream sites in Arizona. *Tree Physiology* 17:151–160.

Körner, Ch. 1994. Scaling from species to vegetation: The usefulness of functional groups. Pp. 117–140 *in* E. D. Schulze and H. A. Mooney, eds. *Biodiversity and ecosystem function*. Springer Verlag, Berlin, Germany.

———. 1995. Towards a better experimental basis for upscaling plant responses to elevated CO_2 and climate warming. *Plant, Cell and Environment* 18:1101–1110.

Küchler, A. W. 1964. *Potential natural vegetation of the coterminous United States*. American Geographical Society Special Publication 36. American Geographical Society, New York.

Kvien, C. S., and W. D. Branch. 1988. Design and use of a fully automated portable rain shelter system. *Agronomy Journal* 80:281–283.

Larsen, B., R. Heitschmidt, and B. Knapp. 1993. Rainout shelter design. PNW93–105. American Society of Agricultural Engineers.

Lauenroth, W. K., D. L. Urban, D. P. Coffin, W. J. Parton, H. H. Shugart, T. B. Kirchner, and T. M. Smith. 1993. Modeling vegetation structure-ecosystem process interactions across sites and ecosystems. *Ecological Modeling* 67:49–80.

Laycock, W. A. 1991. Stable states and thresholds of range condition on North American rangelands: A viewpoint. *Journal of Range Management* 44:427–433.

Le Houerou, H. N., and C. H. Hoste. 1977. Rangeland production and annual rainfall relations in the Mediterranean basin and in the African Shelo-Sudanian zone. *Journal of Range Management* 30:181–189.

Lenihan, J. M., C. Daly, D. Bachelet, and R. P. Neilson. 1998. Simulating broadscale fire severity in a dynamic global vegetation model. *Northwest Science* 72:91–103.

Lentz, D. R., and G. H. Simonson. 1986. *A detailed soils inventory and associated vegetation of Squaw Butte Range Experiment Station.* Special Report 760. Agricultural Experiment Station, Oregon State University, Corvallis.

Leverson, V. 1997. *Potential regional impacts of global warming on precipitation in the western United States.* U.S. Department of the Interior, Bureau of Reclamation Global Climate Change Response Program, Denver.

Lewin, K. F., and L. S. Evans. 1986. Design and performance of an experimental system to determine the effects of rainfall acidity on vegetation. *Transactions of the American Society of Agricultural Engineers* 29:1654–1658.

Lin, G., S. Phillips, and J. R. Ehleringer. 1996. Monsoonal precipitation responses of shrubs in a cold desert community of the Colorado Plateau. *Oecologia* 106:8–17.

Lin, G., and L. Sternberg. 1993. Hydrogen isotopic fractionation by plant roots during water uptake in coastal wetland plants. Pp. 497–510 *in* J. R. Ehleringer, A. E. Hall, and G. D. Farquhar, eds. *Stable isotopes and plant carbon-water relations.* Academic Press, San Diego.

Linton, M. J., J. S. Sperry, and D. G. Williams. 1998. Xylem cavitation in roots and branches of *Juniperus osteosperma* and *Pinus edulis*. *Functional Ecology* 12:906–911.

Little, E. L. 1976. *Atlas of United States trees, volume 3: Minor western hardwoods.* USDA Forest Service, Washington, D.C.

Mahlman, J. D. 1997. Uncertainties in projections of human-caused climate warming. *Science* 278:1416–1417.

Manabe, S., and R. T. Wetherald. 1986. Reduction in summer soil wetness induced by an increase in atmospheric carbon dioxide. *Science* 232:626–628.

Manabe, S., R. T. Wetherald, R. J. Stouffer. 1981. Summer dryness due to an increase of atmospheric CO_2 concentration. *Climatic Change* 3:347–386.

Manogaran, C. 1983. The prairie peninsula: A climate perspective. *Physical Geography* 4:153–166.

Marañón, T., and J. W. Bartolome. 1993. Reciprocal transplants of herbaceous communities between *Quercus agrifolia* woodland and adjacent grassland. *Journal of Ecology* 81:673–682.

Martin, E. C., J. T. Ritchie, S. M. Reese, T. L. Loudon, and B. Knezek. 1988. A large-area, lightweight rainshelter with programmable control. *Transactions of the American Society of Agricultural Engineers* 31:1440–1444.

Martin, P. 1993. Vegetation responses and feedbacks to climate: A review of models and processes. *Climate Dynamics* 8:201–210.

McAuliffe, J. R. 1994. Landscape evolution, soil formation, and ecological patterns and processes in Sonoran Desert bajadas. *Ecological Monographs* 64:111–148.

———. 1995. Landscape evolution, soil formation, and Arizona's desert grasslands. Pp. 100–129 *in* M. P. McClaran and T. R. Van Devender, eds. *The desert grassland.* University of Arizona Press, Tucson.

———. 1998. Rangeland water developments: Conservation solution or illusion? Pp. 310–335 *in Environmental, economic, and legal issues related to rangeland water developments.* Proceedings of a symposium, Nov. 13–15, 1997. Center for the Study of Law, Science and Technology, Arizona State University, Tempe.

———. 1999a. Desert soils. Pp. 87–104 *in* S. Phillips and P. Comus, eds. *The naturalist's guide to the Sonoran Desert.* University of California Press, Berkeley.

———. 1999b. The Sonoran Desert: Landscape complexity and ecological diversity. Pp. 68–114 *in* R. H. Robichaux, ed. *The ecology of Sonoran Desert plants and plant communities.* University of Arizona Press, Tucson.

McAuliffe, J. R., and T. L. Burgess. 1995. Landscape complexity, soil development, and vegetational diversity within a sky island piedmont: A field trip guide to Mt. Lemmon and San Pedro Valley. Pp. 91–108 *in* L. F. DeBano, G. J. Gottfried, R. H. Hamre, C. B. Edminster, P. F. Ffolliott, and A. Ortega-Rubio, tech. coordinators. *Biodiversity and management of the Madrean Archipelago: The sky islands of southwestern United States and Northwestern Mexico.* USDA Forest Service General Technical Report RM-264. Rocky Mountain Forest and Range Experiment Station, Fort Collins, Colo.

McClaran, M. P., and G. R. McPherson. 1995. Can soil organic carbon isotopes be used to describe grass-tree dynamics at a savanna-grassland ecotone and within the savanna? *Journal of Vegetation Science* 6:857–862.

———. 1999. Oak savanna of the American southwest. Pp. 275–287 *in* R. C. Anderson, J. S. Fralish, and J. Baskin, eds. *Savannas, barrens, and rock outcrop plant communities of North America.* Cambridge University Press, Cambridge.

McDonald, E. V., F. B. Pierson, G. N. Flerchinger, and L. D. McFadden. 1996. Application of a process-based soil water balance model to evaluate the influence of late Quaternary climate change on soil-water movement in calcic soils. *Geoderma* 74:167–192.

McPherson, G. R. 1992. Ecology of oak woodlands in Arizona. Pp. 24–33 *in* P. F. Ffolliott, G. J. Gottfried, D. A. Bennett, V. M. Hernandez, C. A. Ortega-Rubio, and R. H. Hamre, tech. coordinators. *Ecology and management of oak and associated woodlands: Perspectives in the southwestern United States and northern Mexico.* USDA Forest Service General Technical Report RM-218. Rocky Mountain Forest and Range Experiment Station, Fort Collins, Colo.

———. 1993. Effects of herbivory and herbs on oak establishment in a semi-arid temperate savanna. *Journal of Vegetation Science* 4:687–692.

———. 1995. The role of fire in the desert grasslands. Pp. 130–151 *in* M. P. McClaran and T. R. Van Devender, eds. *The desert grassland.* University of Arizona Press, Tucson.

———. 1997. *Ecology and management of North American savannas.* University of Arizona Press, Tucson.

McPherson, G. R., T. W. Boutton, and A. J. Midwood. 1993. Stable carbon isotope analysis of soil organic matter illustrates vegetation change at the grassland/woodland boundary in southeastern Arizona, USA. *Oecologia* 93:95–101.

McPherson, G. R., and J. F. Weltzin. 2000. *Disturbance and climate change in United States/Mexico borderland plant communities: A state-of-the-knowledge review.* General Technical Report RMRS-GTR-50. USDA Forest Service, Rocky Mountain Research Station, Fort Collins, Colo.

Mehringer, P. J., and P. E. Wigand. 1990. Comparison of late Holocene environments from woodrat middens and pollen: Diamond Craters, Oregon. Pp. 294–325 *in* J. L. Betancourt, T. R. Van Devender, and P. S. Martin, eds. *Packat middens: The last 40,000 years of biotic change.* University of Arizona Press, Tucson.

Melillo, J. M., A. D. McGuire, D. W. Kicklighter, B. Moore, C. J. Vorosmarty, and A. L. Schloss. 1993. Global climate change and terrestrial net primary production. *Nature* 363:234 240.

Melillo, J. M., I. C. Prentice, G. D. Farquhar, E. D. Schulze, and O.E. Sala. 1996. Terrestrial biotic response to environmental change and feedbacks to climate. Pp. 449–481 *in* J. T. Houghton, L. G. Miera Filho, B. A. Callander, N. Harris, A. Kattenberg, and K. Maskell, eds. 1996. *Climate change 1995: The science of climate change.* Intergovernmental Panel on Climate Change. Cambridge University Press, Cambridge.

Mencuccini, M., and J. C. Comstock. 1997. Vulnerability to cavitation in populations of two desert species, *Hymenoclea salsola* and *Ambrosia dumosa,* from different climatic regions. *Journal of Experimental Botany* 48:1323–1334.

Merriam, C. H. 1890. Results of a biological survey of the San Francisco Mountains region and desert of the Little Colorado, Arizona. *North American Fauna* 3:1–136.

Miller, R. F., and P. E. Wigand. 1994. Holocene changes in semiarid pinyon-juniper woodlands. *BioScience* 44:465–474.

Mitchell J. F. B., T. C. Johns, J. M. Gregory, and S. Tett. 1995. Climate response to increasing levels of greenhouse gases and sulphate aerosols. *Nature* 376:501–504.

Mitchell, V. L. 1976. The regionalization of climate in the western United States. *Journal of Applied Meteorology* 15:920–927.

Monson, R. K., and S. D. Smith. 1982. Seasonal water potential components of Sonoran Desert plants. *Ecology* 63:113–123.

Mooney, H. A., J. Canadell, F. S. Chapin III, J. R. Ehleringer, Ch. Körner, R. E. McMurtrie, W. J. Parton, L. F. Pitelka, and E. D. Schulze. 1999. Ecosystem physiology responses to global change. Pp. 141–189 *in* B. Walker, W. Steffen, J. Canadell, and J. Ingram, eds. *The terrestrial biosphere and global change: Implications for natural and managed systems.* Cambridge University Press, Cambridge.

Morrison, R. B. 1964. *Lake Lahontan: Geology of southern Carson Desert, Nevada.* Geological Survey Professionals Paper 401. U.S. Government Printing Office, Washington, D.C.

Murphy, J. S., and D. D. Briske. 1992. Regulation of tillering by apical dominance: Chronology, interpretive value, and current perspectives. *Journal of Range Management* 45:419–429.

——. 1994. Density-dependent regulation of ramet recruitment by the red:far-red ratio of solar radiation: A field evaluation with the bunchgrass *Schizachyrium scoparium. Oecologia* 97:462–469.

National Assessment Synthesis Team. 2000. *Climate change impacts on the United States: The potential consequences of climate variability and change.* U.S. Global Change Research Program. Cambridge University Press, New York.

Neff, E. L. 1982. Chemical quality and sediment content of runoff water from southeastern Montana rangeland. *Journal of Range Management* 35:130–132.

Neilson, R. P. 1986. High-resolution climatic analysis and southwest biogeography. *Science* 232:27–34.

——. 1987a. Biotic regionalization and climatic controls in western North America. *Vegetatio* 70:135–147.

——. 1987b. On the interface between current ecological studies and the paleobotany of Pinyon-Juniper woodlands. Pp. 93–98 *in* R. L. Everett, ed. *Proceedings, Pinyon-Juniper Conference.* General Technical Report INT-215. USDA Forest Service, Intermountain Research Station, Ogden, Utah.

——. 1991. Climatic constraints and issues of scale controlling regional biomes. Pp. 31–51 *in* M. M. Holland, P. G. Risser, and R. J. Naiman, eds. *Ecotones: The role of landscape boundaries in the management and restoration of changing environments.* Chapman and Hall, New York.

——. 1995. A model for predicting continental-scale vegetation distribution and water balance. *Ecological Applications* 5:362–385.

Neilson, R. P., and R. J. Drapek. 1998. Potentially complex biosphere responses to transient global warming. *Global Change Biology* 4:505–521.

Neilson, R. P., G. A. King, R. L. DeVelice, J. Lenihan, D. Marks, J. Dolph, W. Campbell, and G. Glick. 1989. *Sensitivity of ecological landscapes to global climatic change.* EPA-600-3-89-073, NTIS-PB-90-120-072-AS. U.S. Environmental Protection Agency, Washington, D.C.

Neilson, R. P., G. A. King, and G. Koerper. 1992. Toward a rule-based biome model. *Landscape Ecology* 7:27–43.

Neilson, R. P., and L. H. Wullstein. 1983. Biogeography of two southwest American oaks in relation to atmospheric dynamics. *Journal of Biogeography* 10:275–297.

——. 1985. Comparative drought physiology and biogeography of *Quercus gambelii* and *Quercus turbinella. American Midland Naturalist* 114:259–271.

——. 1986. Microhabitat affinities of Gambel oak seedlings. *Great Basin Naturalist* 46:294–298.

Nelson, E. W. 1934. *The influence of grazing and precipitation upon black grama grass range.* USDA Technical Bulletin 409. Washington, D.C.

Nesmith, D. S., A. Miller, and J. T. Ritchie. 1990. An irrigation system for plots under a rain shelter. *Agricultural Water Management* 17:409–414.

Nesmith, D. S., and J. T. Richie. 1992. Short- and long-term responses of corn to a pre-anthesis soil water deficit. *Agronomy Journal* 84:107–113.

Nicks, A. D., and L. J. Lane. 1989. Weather generator. Pp. 2.1–2.19 *in* L. J. Lane and M. A. Nearing, eds. USDA water erosion prediction project: Hillslope profile model documentation. NSERL Report No. 2. USDA–ARS National Soil Erosion Research Laboratory, West Lafayette, Ind.

NOAA. 1996. *Climatological data annual summary: Arizona.* Department of Commerce, Asheville, N.C.

Nobel, P. S. 1994. Root-soil responses to water pulses in dry environments. Pp. 285–304 *in* M. M. Caldwell and R. W. Pearcy, eds. *Exploitation of environmental heterogeneity by plants: Ecophysiological processes above and below ground.* Academic Press, San Diego.

Nobel, P. S., D. M. Alm, and J. Cavelier. 1992. Growth respiration, maintenance respiration, and structural-carbon costs for roots of three desert succulents. *Functional Ecology* 6:79–85.

Nobel, P. S., and M. Cui. 1992a. Hydraulic conductances of the soil, the root-soil air gap, and the root: Changes for desert succulents in drying soil. *Journal of Experimental Botany* 43:319–326.

——. 1992b. Shrinkage of attached roots of *Opuntia ficus-indica* in response to lowered water potentials: Predicted consequences for water uptake or loss to soil. *Annals of Botany* 70:485–491.

Norwine, J. 1978. Twentieth-century semi-arid climates and climatic fluctuations in Texas and northeastern Mexico. *Journal of Arid Environments* 1:313–325.

Nowak, C. L., R. S. Nowak, R. J. Tausch, and P. E. Wigand. 1994. Tree and shrub dynamics in northwestern Great Basin woodland and shrub steppe during the late-Pleistocene and Holocene. *American Journal of Botany* 81:265–277.

Noy-Meir, I. 1973. Desert ecosystems: Environment and producers. *Annual Review of Ecology and Systematics* 4:25–51.

Noy-Meir, E. 1979. Structure and function of desert ecosystems. *Israel Journal of Botany* 28:1–19.

Nyandiga, C. O., and G. R. McPherson. 1992. Germination of two warm-temperate oaks, *Quercus emoryi* and *Quercus arizonica. Canadian Journal of Forest Research* 22:1395–1401.

Office of Science and Technology Policy. 1996. *Climate change.* U.S. Government Printing Office, Washington, D.C.

Ojima, D. S., B. O. M. Dirks, E. P. Glenn, C. E. Owensby, and J. O. Scurlock. 1993. Assessment of C budget for grasslands and drylands of the world. *Water, Air, and Soil Pollution* 70:95–109.

Olson, B. E., and J. H. Richards. 1988. Annual replacement of the tillers of *Agropyron desertorum* following grazing. *Oecologia* 76:1–6.

Olson, K. C., R. S. White, and B. W. Sindelar. 1985. Responses of vegetation of the northern great plains to precipitation amount and grazing intensity. *Journal of Range Management* 38:357–361.

Opperman, D. P. J., J. J. Human, and M. F. Vijoen. 1977. Evapotranspirasiestudies op *Themeda triandra* Forsk onder veldtoestande. P. 342 *in* B. J. Huntley and B. H. Walker, eds. 1982. *Ecology of tropical savannas.* Springer-Verlag, Berlin.

Osborn, H. B. 1983. *Precipitation characteristics affecting responses of southwestern rangelands.* Agricultural Reviews and Manuals ARM-W-34. USDA Agricultural Research Service (Western Region), Oakland, Calif.

Overpeck, J. T. 1996. Warm climate surprises. *Science* 271:1820–1821.

Overpeck, J. T., P. J. Bartlein, and T. Webb, III. 1991. Potential magnitude of future vegetation change in eastern North America: Comparisons with the past. *Science* 254:692–695.

Owens, L. B., W. M. Edwards, and R. W. Van Kwuren. 1983. Surface runoff water quality comparisons between unimproved pasture and woodland. *Journal of Environmental Quality* 12:518–522.

Owensby, C. E., P. I. Coyne, J. M. Ham, L. A. Auen, and A. K. Knapp. 1993. Biomass production in a tallgrass prairie ecosystem exposed to ambient and elevated CO_2. *Ecological Applications* 3:644–653.

Parker, L. W., P. F. Santos, J. Phillips, and W. G. Whitford. 1984. Carbon and nitrogen dynamics during the decomposition of litter and roots of a Chihuahuan desert annual, *Lepidium lasiocarpum. Ecological Monographs* 54:339–360.

Parton, W. J., J. M. O. Scurlock, D. S. Ojima, D. S. Schimel, D. O. Hall, M. B. Coughenour, E. M. Garcia, T. G. Gilmanov, A. Kamnalrut, J. I. Kinyamario, T. Kirchner, S. P. Long, J. C. Menaut, O. E. Sala, R. J. Scholes, J. A. Van Veen. 1995. Impact of climate change on grassland production and soil carbon worldwide. *Global Change Biology* 1:13–22.

Parton, W. J., J. W. B. Stewart, and C. V. Cole. 1988. Dynamics of C, N, P, and S in grassland soils: A model. *Biogeochemistry* 5:109–131.

Pase, C. P. 1969. Survival of *Quercus turbinella* and *Quercus emoryi* in an Arizona chaparral community. *Southwestern Naturalist* 14:149–156.

Pase, C. P, and D. E. Brown. 1982. Interior chaparral. *Desert Plants* 4:95–99.

Pase, C. P., and R. R. Johnson. 1968. *Flora and vegetation of the Sierra Ancha Experimental Forest, Arizona.* USDA Forest Service Research Paper RM-41. Rocky Mountain Forest and Range Experiment Station, Fort Collins, Colo.

Passey, H. B., V. K. Hugie, E. W. Williams, and D. E. Ball. 1982. *Relationships between soil, plant community, and climate on rangelands of the intermountain west.* USDA Technical Bulletin Number 1669. Soil Conservation Service, Washington, D.C.

Passioura, J. B. 1988. Water transport in and to roots. *Annual Review of Plant Physiology and Plant Molecular Biology* 39:245–265.

Pastor, J., and W. M. Post. 1988. Response of northern forests to CO_2-induced climate change. *Nature* 334:55–58.

Peco, B., and T. Espigares. 1994. Floristic fluctuations in annual pastures: The role of competition at the regenerative stage. *Journal of Vegetation Science* 5:457–462.

Peet, R. K. 1974. The measurement of species diversity. *Annual Review of Ecology and Systematics* 5:285–307.

Peláez, D. V., R. A. Distel, R. M. Bóo, O. R. Elia, and M. D. Mayor. 1994. Water relations between shrubs and grasses in semi-arid Argentina. *Journal of Arid Environments* 27:71–78.

Peterson, F. F. 1981. *Landforms of the basin and range defined for soil survey.* Nevada Agricultural Experiment Station Technical Bulletin 28. University of Nevada, Reno.

Peterson, G., C. R. Allen, and C. S. Holling. 1998. Ecological resilience, biodiversity, and scale. *Ecosystems* 1:6–18.

Phillips, W. S. 1963. Depth of roots in soil. *Ecology* 44:424.

Pitelka, L. R., and the Plant Migration Workshop Group. 1997. Plant migration and climate change. *American Scientist* 85:464–472.

Platt, J. R. 1964. Strong inference. *Science* 146:347–353.

Pockman, W. T., and J. S. Sperry. 2000. Vulnerability to xylem cavitation and the distribution of Sonoran Desert vegetation. *American Journal of Botany* 87:1287–1299.

Polley, H. W., H. B. Johnson, H. S. Mayeux, and C. R. Tischler. 1996. Are some of the recent changes in grassland communities a response to rising CO_2 concentrations? Pp. 177–195 *in* Ch. Körner and F. A. Bazzaz, eds. *Carbon dioxide, populations, and communities.* Academic Press, San Diego.

Polley, H. W., H. S. Mayeux, H. B. Johnson, and C. R. Tischler. 1997. Viewpoint: Atmospheric CO_2, soil water, and shrub/grass ratios on rangelands. *Journal of Range Management* 50:278–284.

Poorter, H. 1993. Interspecific variation in the growth response of plants to an elevated ambient CO_2 concentration. *Vegetatio* 104/105:77–97.

Pregitzer, K. S., R. L. Hendrick, and R. Fogel. 1993. The demography of fine roots in response to patches of water and nitrogen. *New Phytologist* 125:575–580.

Pregitzer, K. S., D. R. Zak, J. Maziasz, J. DeForest, P. S. Curtis, and J. Lussenhop. 2000. Interactive effects of atmospheric CO_2 and soil-N availability on fine roots of *Populus tremuloides*. *Ecological Applications* 10:18–33.

Radford, P. J. 1967. Growth analysis formulae: Their use and abuse. *Crop Science* 7:171–175.

Reece, P. E., R. P. Bode, and S. S. Waller. 1988. Vigor of needle-and-thread and blue grama after short duration grazing. *Journal of Range Management* 41:287–291.

Reed, M. J., and R. A. Peterson. 1961. *Vegetation, soils and cattle response to grazing on northern Great Plains.* USDA Forest Service Technical Bulletin 1952. Washington, D.C.

Reynolds, J. F., P. R. Kemp, and J. D. Tenhunen. 2000. Effects of long-term rainfall variability on evapotranspiration and soil water distribution in the Chihuahuan Desert: A modeling analysis. *Plant Ecology* 150:145–159.

Reynolds, J. F., R. A. Virginia, P. R. Kemp, A. G. de Soyza, and D. C. Tremmel. 1999. Impact of drought on desert shrubs: Effects of seasonality and degree of resource island development. *Ecological Monographs* 69:69–106.

Rice, C. W., T. C. Todd, J. M. Blair, T. R. Seastedt, R. A. Ramundo, and G. W. T. Wilson. 1998. Belowground biology and processes. Pp. 244–265 *in* A. K. Knapp, J. M. Briggs, D. C. Hartnett, S. L. Collins, eds. *Grassland dynamics: Long-term ecological research in tallgrass prairie.* Oxford University Press, New York.

Rice, K. J. 1989. Impacts of seed banks on grassland community structure and population dynamics. Pp. 211–230 *in* M. Leck, V. T. Parker, and R. L. Simpson, eds. *Ecology of soil seedbanks.* Academic Press, San Diego.

Richards, J. H. 1984. Root response to defoliation in two *Agropyron* bunchgrasses: Field observations with an improved root periscope. *Oecologia* 64:21–25.

Richards, J. H., and M. M. Caldwell. 1987. Hydraulic lift: Substantial nocturnal water transport between soil layers by *Artemisia tridentata* roots. *Oecologia* 73:486–489.

Ries, R. E., and L. G. Zachmeier. 1985. Automated rainout shelter for controlled water research. *Journal of Range Management* 38:353–357.

Rind, D., R. Goldberg, J. Hansen, C. Rosenzweig, and R. Ruedy. 1990. Potential evapotranspiration and the likelihood of future drought. *Journal of Geophysical Research* 95(D7):9983–10004.

Rosenzweig, M. L. 1968. Net primary productivity of terrestrial communities: Prediction from climatological data. *American Naturalist* 102:67–74.

Sala, O. E., R. A. Golluscio, W. K. Lauenroth, and A. Soriano. 1989. Resource partitioning between shrubs and grasses in the Patagonian steppe. *Oecologia* 81:501–505.

Sala, O. E., and W. K. Lauenroth. 1982. Small rainfall events: An ecological role in semiarid regions. *Oecologia* 53:301–304.

———. 1985. Root profiles and the ecological effect of light rainshowers in arid and semiarid regions. *American Midland Naturalist* 114:406–408.

Sala, O. E., W. K. Lauenroth, and R. A. Golluscio. 1997. Plant functional types in temperate semi-arid areas. Pp. 217–233 *in* T. M. Smith, H. H. Shugart, and F. I. Woodward, eds. *Plant functional types.* Cambridge University Press, Cambridge.

Sala, O. E., W. J. Parton, L. A. Joyce, and W. K. Lauenroth. 1988. Primary production of the central grassland region of the United States. *Ecology* 69:40–45.

Samson, F., and F. Knopf. 1994. Prairie conservation in North America. *Bioscience* 44:418–421.

Sanchini, P. J. 1981. Population structure and fecundity patterns in *Quercus emoryi* and *Quercus arizonica* in southeastern Arizona. Ph.D. diss., University of Colorado, Boulder.

SAS Institute. 1988. *Users guide, release 6.03 edition.* SAS Institute, Cary, N.C.

Schaeffer, S. M., D. G. Williams, and D. C. Goodrich. 2000. Transpiration in cottonwood/willow forest patches estimated from sap flux. *Agricultural and Forest Meteorology* 195:257–270.

Schimel, D. S. 1993. Population and community processes in the response of terrestrial ecosystems to global change. Pp. 45–56 *in* P. M. Kareiva, J. G. Kingsolver, R. B. Huey, eds. *Biotic interactions and global change.* Sinauer Associates, Sunderland, Mass.

Schimel, D., J. Melillo, H. Tian, A. D. McGuire, D. Kicklighter, T. Kittel, N. Rosenbloom, S. Running, P. Thornton, D. Ojima, W. Parton, R. Kelly, M. Sykes, R. Neilson, and B. Rizzo. 2000. Contribution of increasing CO_2 and climate to carbon storage by ecosystems in the United States. *Science* 287:2004–2006.

Schlesinger, W. H., P. J. Fonteyn, and G. M. Marion. 1987. Soil moisture content and plant transpiration in the Chihuahuan Desert of New Mexico. *Journal of Arid Environments* 12:119–126.

Schlesinger, W. H., J. F. Reynolds, G. L. Cunningham, L. F. Huenneke, W. M. Jarrell, R. A. Virginia, and W. G. Whitford. 1990. Biological feedbacks in global desertification. *Science* 247:1043–1048.

Schneider, S. H. 1993. Scenarios of global warming. Pp. 9–23 *in* P. M. Kareiva, J. G. Kingsolver, and R. B. Huey, eds. *Biotic interactions and global change.* Sinauer Associates, Sunderland, Mass.

Scholander, P. F., H. T. Hammel, E. D. Bradstreet, and E. A. Hemmingsen. 1965. Sap pressure in vascular plants. *Science* 148:339–346.

Scholes, R. J. 1993. Nutrient cycling in semi-arid grasslands and savannas: its influence on pattern, productivity, and stability. Pp. 1331–1350 *in Proceedings of the XVII International Grassland Congress, February 8–21, 1993.* Palmerstown North, New Zealand.

Scholes, R. J., and S. R. Archer. 1997. Tree-grass interactions in savannas. *Annual Review of Ecology and Systematics* 28:517–544.

Schulze, E. D., H. A. Mooney, O. E. Sala, E. Jobbagy, N. Buchmann, G. Bauer, J. Canadell, R. B. Jackson, J. Loreti, M. Oesterheld, and J. R. Ehleringer. 1996. Rooting depth, water availability, and vegetation cover along an aridity gradient in Patagonia. *Oecologia* 108:503–511.

Schupp, E. W. 1995. Seed-seedling conflicts, habitat choice, and patterns of plant recruitment. *American Jounal of Botany* 82:399–409.

Schwinning, S., and J. R. Ehleringer. 2001. Water use trade-offs and optimal adaptations to pulse-driven arid ecosytems. *Journal of Ecology* 89:464–480.

Scott, R. L., W. J. Shuttleworth, T. O. Keefer, and A. W. Warrick. 2000. Modeling multiyear observations of soil moisture recharge in the semiarid American Southwest. *Water Resouces Research* 36:2233–2248.

Sellers, P. J., R. E. Dickinson, D. A. Randall, A. K. Betts, F. G. Hall, J. A. Berry, G. J. Collatz, A. S. Denning, H. A. Mooney, C. A. Nobre, N. Sato, C. B. Field, and A. Henderson-Sellers. 1997. Modelling the exchanges of energy, water, and carbon between continents and the atmosphere. *Science* 275:502–509.

Shannon, C. E., and W. Weaver. 1949. *The mathematical theory of communication.* University of Illinois Press, Urbana.

Silvertown, J., M. E. Dodd, K. McConway, J. Potts, and M. Crawley. 1994. Rainfall, biomass variation, and community composition in the Park Grass Experiment. *Ecology* 75:2430–2437.

Simpson, E. H. 1949. Measurement of diversity. *Nature* 163:688.

Simpson, R. L., M. A. Leck, and V. T. Parker. 1989. Seed banks: General concepts and methodological issues. Pp. 3–8 *in* M. A. Leck, V. T. Parker, and R. L. Simpson, eds. *Ecology of soil seed banks.* Academic Press, San Diego.

Sims, P. L., and J. S. Singh. 1978. The structure and function of ten western North American grasslands. III. Net primary production, turnover, and efficiencies of energy capture and water use. *Journal of Ecology* 66:547–597.

Singh, J. S., W. K. Lauenroth, R. K. Heitschmidt, and J. L. Dodd. 1982. Structural and functional attributes of the vegetation of northern mixed prairie of North America. *Botanical Review* 49:117–149.

Smith, M. D., and A. K. Knapp. 1999. Exotic plant species in a C_4- dominated grassland: Invasibility, disturbance, and community structure. *Oecologia* 120:605–612.

Smith, S. D., T. E. Huxman, S. F. Zitzer, T. N. Charlet, D. C. Housman, J. S.

Coleman, L. K. Fenstermaker, J. R. Seemann, and R. S. Nowak. 2000. Elevated CO_2 increases productivity and invasive species success in an arid ecosystem. *Nature* 408:79–82.

Smith, S. D., A. B. Wellington, J. A. Nachlinger, and C. A. Fox. 1991. Functional responses of riparian vegetation to streamflow diversion in the eastern Sierra Nevada. *Ecological Applications* 1:89–97.

Snyder, K. A., and D. G. Williams. 2000. Water sources used by riparian trees varies among stream types on the San Pedro River, Arizona. *Agricultural and Forest Meterology* 105:227–240.

Snyder, K. A., D. G. Williams, and V. L. Gempko. 1998. Water source determination for cottonwood, willow, and mesquite in riparian forest stands. Pp. 185–188 *in* E. F. Wood, A. G. Chebouni, D. C. Goodrich, D. J. Seo, and J. R. Zimmerman, tech. coordinators. *Proceedings from the Special Symposium on Hydrology.* American Meteorological Society, Boston.

Soil Science Society of America. 1997. *Glossary of soil science terms 1996.* Soil Science Society of America, Madison, Wisc.

Southwood, T.R.E. 1978. *Ecological methods.* Chapman and Hall, London.

Sperry, J. S., F. R. Alder, G. S. Campbell, and J. P. Comstock. 1998. Limitation of plant water use by rhizosphere and xylem conductance: Results from a model. *Plant, Cell and Environment* 21:347–359.

Stephenson, N. L. 1990. Climatic control of vegetation distribution: The role of the water balance. *American Naturalist* 135:649–670.

Sternberg, P. D., M. A. Anderson, R. C. Graham, J. L. Beyers, and K. R. Tice. 1996. Root distribution and seasonal water status in weathered granitic bedrock under chaparral. *Geoderma* 72:89–98.

Steudle, E., and C. A. Peterson. 1998. How does water get through roots? *Journal of Experimental Botany* 49:775–788.

Stout, D. O., A. McLean, B. Brooke, and J. Hall. 1980. Influences of simulated grazing (clipping) on pinegrass growth. *Journal of Range Management* 33:286–291.

Svejcar, T., R. Angell, and R. Miller. 1999. Fixed location rainout shelters for studying precipitation effects on rangelands. *Journal of Arid Environments* 42:187–193.

Takahashi, H. 1994. Hydrotropism and its interaction with gravitropism in roots. *Plant and Soil* 165:301–308.

Tang, M., and E. R. Reiter. 1984. Plateau monsoons of the Northern Hemisphere: A comparison between North America and Tibet. *Monthly Weather Review* 112:617–637.

Tausch, R. J., P. E. Wigand, and W. Burkhardt. 1993. Viewpoint: Plant community thresholds, multiple steady states, and multiple successional pathways: Legacy of the Quaternary? *Journal of Range Management* 46:439–447.

Teare, I. D., H. Schimmelpfennig, and R. P. Waldren. 1973. Rainout shelter and drainage lysimeters to quantitatively measure drought stress. *Agronomy Journal* 65:544–547.

Thompson, R. S. 1990. Late Quaternary vegetation and climate in the Great Basin. Pp. 352–359 *in* J. L. Betancourt, T. R. Devender, and P. S. Martin, eds. *Packrat middens: The last 40,000 years.* University of Arizona Press, Tucson.

Thurow, T. L. 1991. Hydrology and erosion. Pp. 141–159 *in* R. K. Heitschmidt and J. W. Stuth, eds. *Grazing management: An ecological perspective.* Timber Press, Portland, Oreg.

Thurow, T. L., W. H. Blackburn, and C. A. Taylor, Jr. 1988. Infiltration and interrill erosion responses to selected livestock grazing strategies, Edwards Plateau, Texas. *Journal of Range Management* 41:296–302.

Tilman, D. 1993. Species richness of experimental productivity gradients: How important is colonization limitation? *Ecology* 74:2179–2191.

Tilman, D., J. Knops, D. Wedin, P. Reich, M. Ritchie, and E. Siemann. 1997. The influence of functional diversity and composition on ecosystem processes. *Science* 277:1300–1302.

Topp, G. C., and J. L. Davis. 1985. Measurement of soil water content using time domain reflectometry TDR: A field evaluation. *Soil Science Society of America Journal* 49:19–24.

Topp, G. C., J. L. Davis, and A. P. Annan. 1980. Electromagnetic determination of soil water content: Measurement in coaxial transmission lines. *Water Resources Research* 16:574–582.

Trettin, C. C., D. W. Johnson, and D. E. Todd, Jr. 1999. Forest nutrient and carbon pools at Walker Branch Watershed: Changes over a 21-year period. *Soil Science Society of America Journal* 63:1436–1448.

Tyree, M. T., and J. D. Alexander. 1993. Plant water relations and the effects of elevated CO_2: A review and suggestions for future research. *Oecologia* 89:580–587.

Upchurch, D. R., J. T. Richie, and M. A. Foale. 1983. Design of a large dual-structure rainout shelter. *Agronomy Journal* 75:845–848.

USDA. 1997. *National range and pasture handbook.* USDA Natural Resources Conservation Service, Washington, D.C.

Valentini, R., G.E.S. Mugnozza, and J. R. Ehleringer. 1992. Hydrogen and carbon isotope ratios of selected species of a Mediterranean macchia ecosystem. *Functional Ecology* 6:627–631.

Vavra, M., W. A. Laycock, and R. D. Pieper, eds. 1994. *Ecological implications of livestock herbivory in the West.* Society for Range Management, Denver, Colo.

VEMAP Members. 1995. Vegetation/ecosystem modeling and analysis project: Comparing biogeography and biogeochemistry models in a continental-scale study of terrestrial ecosytem responses to climate change and CO_2 doubling. *Global Biogeochemical Cycles* 9:407–437.

Vitousek, P. M. 1994. Beyond global warming: Ecology and global change. *Ecology* 75:1861–1876.

Voorhees, W. B. 1989. Root activity related to shallow and deep compaction. Pp. 173–186 *in* W. E. Larsen, ed. *Mechanics and related processes in structured agricultural soils.* Kluwer Academic Press, Dordrecht, Boston.

Waggoner, P. E. 1989. Anticipating the frequency distribution of precipitation if climate alters its mean. *Agricultural and Forest Meteorology* 47:321–337.

Walker, B. H. 1996. Predicting a future terrestrial biosphere: Challenges to GCTE science. Pp. 595–607 *in* B. Walker and W. Steffen. *Global change and terrestrial ecosystems.* Cambridge University Press, Cambridge.

Walter, H. 1954. Die verbuschung, eine erscheinung der subtropischen savannengebiete, und ihre ökologischen ursachen. *Vegetatio* 5/6:6–10.

——. 1971. *Ecology of tropical and subtropical vegetation.* Oliver & Boyd, Edinburgh.

——. 1979. *Vegetation of the earth and ecological systems of the geo-biosphere,* 2nd ed. Springer-Verlag, New York.

Wan, C., R. E. Sosebee, and B. L. McMichael. 1994. Hydraulic properties of shallow vs. deep lateral roots in a semiarid shrub, *Gutierrezia sarothrae. American Midland Naturalist* 131:120–127.

Warner, R. R., and P. L. Chesson. 1985. Coexistence mediated by recruitment fluctuations: A field guide to the storage effect. *American Naturalist* 125:769–787.

Watson, R. T., M. C. Zinyowera, R. H. Moss, and D. J. Dokken. 1997. *The regional impacts of climate change: An assessment of vulnerability.* Intergovernmental Panel on Climate Change Working Group II. Cambridge University Press, Cambridge.

Weaver, J. E. 1958. Classification of root systems of forbs of grassland and a consideration of their significance. *Ecology* 39:393–401.

——. 1968. *Prairie plants and their environment: A fifty-year study in the Midwest.* University of Nebraska, Lincoln

Webb, W. L., W. K. Lauenroth, S. T. Szarek, and R. S. Kinerson. 1983. Primary production and abiotic controls in forests, grasslands, and desert. *Ecology* 64:134–151.

Webb. W., S. Szarek, W. K. Laurenroth, and R. Kinerson. 1978. Primary production and water use in native forest, grassland, and desert ecosystems. *Ecology* 59:1239–1247.

Wells, P. V. 1970. Postglacial vegetation history of the Great Plains. *Science* 167:1574–1582.

Weltz, M. A. 1987. Observed and estimated water budgets for south Texas range-lands. Ph.D. diss., Texas A&M University, College Station.

Weltzin, J. F., and G. R. McPherson. 1995. Potential effects of climate change on lower treelines in the southwestern United States. Pp. 180–193 *in* L. F. DeBano, G. J. Gottfried, R. H. Hamre, C. B. Edminster, P. F. Ffolliott, and A. Ortega-Rubio, tech. coordinators. *Biodiversity and management of the Madrean Archipelago: The sky islands of southwestern United States and Northwestern Mexico.* USDA Forest Service General Technical Report RM-264. Rocky Mountain Forest and Range Experiment Station, Fort Collins, Colo.

———. 1997. Spatial and temporal soil moisture resource partitioning by trees and grasses in a temperate savanna, Arizona, USA. *Oecologia* 112:156–164.

———. 1999. Facilitation of conspecific seedling recruitment and shifts in temperate savanna ecotones. *Ecological Monographs* 69:513–534.

———. 2000. Implications of precipitation redistribution for shifts in temperate savanna ecotones. *Ecology* 81:1902–1903.

Wendland, W. M., and R. A. Bryson. 1981. Northern hemisphere airstream regions. *Monthly Weather Review* 109:255–270.

Wessman, C. A., J. D. Aber, D. L. Peterson, and J. M. Melillo. 1988. Foliar analysis using near infrared reflectance spectroscopy. *Canadian Journal of Forest Research* 18:6–11.

Westoby, M., B. Walker, and I. Noy-Meir. 1989. Range management on the basis of a model which does not seek to establish equilibrium. *Journal of Arid Environments* 17:235–239.

Whitford, W. G., and R. K. Fenton. 1999. Biopedturbation by mammals in deserts: A review. *Journal of Arid Environments* 41:203–230.

Whittaker, R. H., and W. A. Niering. 1965. Vegetation of the Santa Catalina Mountains, Arizona. II. A gradient analysis of the south slope. *Ecology* 46:429–452.

Wigand, P. E., and C. L. Nowak. 1992. Dynamics of northwest Nevada plant communities during the last 30,000 years. Pp. 40–62 *in* C. A. Hall, V. Doyle-Jones, and B. Widawski, eds. *The history of water: Eastern Sierra Nevada, Owens Valley, White-Inyo Mountains,* vol. 4. White Mountain Research Station Symposium, Los Angeles.

Williams, D. G., J. P. Brunel, S. M. Schaeffer, and K. A. Snyder. 1998. Biotic controls over the functioning of desert riparian ecosystems. Pp. 43–48 *in* E. F. Wood, A. G. Chebouni, D. C. Goodrich, D. J. Seo, and J. R. Zimmerman, tech. coordinators. *Proceedings from the Special Symposium on Hydrology.* American Meteorological Society, Boston.

Williams, D. G., and J. R. Ehleringer. 2000. Intra- and interspecific variation for summer precipitation use in pinyon-juniper woodlands. *Ecological Monographs* 70:517–537.

Williams, K., F. Percival, J. Merino, and H. A. Mooney. 1987. Estimation of tissue construction costs from heat of combustion and organic nitrogen content. *Plant, Cell and Environment* 10:725–734.

Wilson, J. B., and A.D.Q. Agnew. 1992. Positive-feedback switches in plant communities. *Advances in Ecological Research* 23:263–336.

Wilson, K. B., D. D. Baldocchi, and P. J. Hanson. 2000. Quantifying stomatal and non-stomatal limitations to carbon assimilation resulting from leaf aging and drought in mature deciduous tree species. *Tree Physiology* 20:787–797.

Wittwer, S. H. 1995. *Food, climate, and carbon dioxide: The global environment and world food production.* CRC Press, Boca Raton, Fla.

Woodward, F. I. 1987. *Climate and plant distribution.* Cambridge University Press, London.

Wrage, K. L., F. R. Gartner, and J. L. Butler. 1994. Inexpensive rain gauges constructed from recyclable 2-liter plastic soft drink bottles. *Journal of Range Management* 47:249–250.

Yoder, C. K., T. W. Boutton, T. L. Thurow, and A. J. Midwood. 1998. Differences in soil water use by annual broomweed and grasses. *Journal of Range Management* 51:200–206.

Zar, J. H. 1996. *Biostatistical analysis.* Prentice Hall, Upper Saddle River, N.J.

Zhang, J., and J. T. Romo. 1995. Impacts of defoliation on tiller production and survival in northern wheatgrass. *Journal of Range Management* 48:115–120.

Zobel, R. W. 1991. Root growth and development. Pp. 61–71 *in* D. Keister and P. Cregan, eds. *Symposium proceedings on rhizosphere and plant growth, May 8–11, 1989.* Beltsville Agricultural Research Center (BARC), Beltsville, Md. Kluwer, Boston.

About the Editors

Jake F. Weltzin is an assistant professor of ecology and evolutionary biology at the University of Tennessee. His research is focused on the response of plant populations, plant and animal communities, and ecosystems to anthropogenic climate change. He uses large-scale field experiments to investigate the response of terrestrial ecosystems to changes in the amount and distribution of precipitation, global warming, and increasing concentrations of carbon dioxide in the atmosphere. His research is published in regional, national, and international ecological journals. This is his first book.

Guy R. McPherson is a professor in the School of Renewable Natural Resources and the Department of Ecology and Evolutionary Biology at the University of Arizona. His teaching and research are focused on the development and creative application of ecological theory within the context of biological conservation. His research has been published in most major journals that deal with ecology and management of natural resources, and has been synthesized in two books: *Ecology and Management of North American Savannas* (University of Arizona Press, 1997) and *Applying Ecology Toward Natural Resource Management* (with Steve DeStefano, Cambridge University Press, in press).

About the Contributors

Raymond Angell is a rangeland scientist working with the U.S. Department of Agriculture–Agricultural Research Service (USDA–ARS). He is stationed at the Eastern Oregon Agricultural Research Center in Burns, Oregon. Dr. Angell obtained a B.S. in wildlife biology and an M.S. in agronomy from Kansas State University. He received a Ph.D. in range science from Texas A&M University, where his research focused on prescribed burning of gulf cordgrass as a management tool for improving livestock nutrition during winter. Presently his research interests include the effect of climate on rangeland plants, management of native meadow vegetation, livestock grazing and livestock/plant interactions, and the dynamics of carbon dioxide flux over burned and unburned rangelands. His most recent publication concerned the carbon dioxide flux over unburned sagebrush rangeland.

Jon Bates is a rangeland scientist with the USDA–ARS in Burns, Oregon. He received a B.S. in agricultural business at Cornell University, an M.S. in agricultural economics at Oregon State University, and a Ph.D. in rangeland resources at Oregon State University. His present research focuses on the effects of climate on rangeland plants, restoration of sagebrush steppe grasslands, aridland nitrogen cycling, native seed production, and prescription grazing after fire and juniper cutting.

John M. Blair is a professor of biology at Kansas State University, and principal investigator in the Konza Prairie Long-Term Ecological Research Program. His research interests include grassland ecosystem ecology, with an emphasis on ecological processes in soils and responses to natural and anthropogenic disturbances. His research is directed at understanding the biological and physical factors controlling ecosystem processes, and the relationship of soil/plant/litter interactions to ecosystem function. His current studies focus on the effects of fire on soil nitrogen cycling and plant responses, and the impacts of climate change on belowground processes.

James H. Brown is a distinguished professor of biology at the University of New Mexico. His voluminous research is driven by a curiosity about the diversity of life. Over most of his career he has simultaneously conducted two major research programs: a long-term project in experimental desert ecology, manipulating and monitoring a small patch of Chihuahuan Desert near Portal, Arizona; and, in macroecology, compiling large data sets on body size, abundance, area of geographic range, and other attributes of many species of birds, mammals, and other taxa. Recent research has focused on patterns related to body size and the search for fundamental mechanistic principles that produce or constrain such allometric relationships. Dr. Brown was recently awarded the Eugene P. Odum Award for Excellence in Ecology Education and the George Mercer Award for an outstanding paper by a younger investigator (shared with B. J. Enquist, G. W. West, and E. L. Charnov). He has edited 3 books, authored or coauthored 3 books, and published 145 papers and 28 chapters in edited volumes, among other accomplishments.

Jonathan D. Carlisle is a research assistant in the Division of Biology at Kansas State University. His primary interest is plant responses to anthropogenic stress. His technical expertise lies in micro-meteorology, instrumentation, and plant ultrastructure in ecosystems from transitional boreal forest to mesic grasslands.

Brett T. Danner is an environmental scientist with Burns & McDonnell Engineering, Inc. in Kansas City, Missouri. As a graduate student at Kansas State University, he focused on the mechanisms of tree seedling establishment in tallgrass prairie. Results from his thesis research have been published in several ecological journals.

Phil A. Fay is a research assistant professor of biology at Kansas State University. His current research focuses on anthropogenic climatic change impacts on grassland communities and ecosystems. His work combines laboratory, greenhouse, and large-scale field manipulations, all aimed at understanding the role of climatic variability, especially extreme climatic events, on the structure and function of tallgrass prairie ecosystems. His research expertise spans several areas, including plant ecophysiology, population ecology, and plant-animal interactions. His work has been funded by the National Science Foundation, USDA, and Department of Energy and published in numerous ecological journals.

Marshall R. Haferkamp has worked in the field of range management for over 35 years. He received degrees in range management from Colorado State University (B.S. in 1966 and M.S. in 1968) and the University of Arizona (Ph.D. in 1975). He has studied soil-plant-environment-herbivore interactions on native ranges and seeded pastures in several ecoregions in the United States while working for South Dakota State University, University of Arizona, Texas A&M University, Oregon State University, and the USDA–ARS in Burns, Oregon, and Miles City, Montana. He has taught classes in range plants, range ecology, and range restoration. He worked for many years in range restoration in Arizona, Texas, Oregon, and Montana. More recently he has worked to quantify Japanese brome's invasive potential, competitive impact, and contribution to forage and livestock production in the Northern Great Plains. He has also conducted research to determine the role some Northern Great Plains' ranges play in regulating atmospheric carbon dioxide flux. His research has resulted in 119 publications, including 40 refereed technical journal articles and 33 presentations at national and international professional meetings and symposia.

Paul J. Hanson is a senior research staff member of the Environmental Sciences Division, Oak Ridge National Laboratory, Oak Ridge, Tennessee. He received M.S. and Ph.D. degrees from the University of Minnesota in the fields of plant and forest tree physiology in 1983 and 1986, respectively. Dr. Hanson's current research focuses on the impacts of climatic change on the physiology, growth, and biogeochemical cycles of eastern deciduous forest ecosystems. He has also conducted work on the impacts of air pollutants on plants, the deposition of gaseous nitrogen compounds to plant surfaces, and the exchange of mercury vapor between terrestrial surfaces and the atmosphere. Dr. Hanson has authored or coauthored over 70 peer-reviewed publications and served as an associate editor of the *Journal of Environmental Quality* for six years, from 1995 through 2000.

Rodney K. Heitschmidt is a rangeland ecologist and research leader and superintendent of the USDA–ARS's Fort Keogh Livestock and Range Research Laboratory near Miles City, Montana. Previously, he was a professor of rangeland ecology and management with the Texas Agricultural Experiment Station near Vernon, Texas, where he served as research coordinator at the Texas Experimental Ranch. His research interests center on developing an understanding of how various grazing tactics, in concert with climate, alter function and structure of rangeland ecosystems. His areas of greatest interest include understanding the interaction effects of grazing and climatic conditions on plant and animal growth dynamics, primary and secondary productivity, plant species composition, water quality and yield, sediment production, and energy flow. He has published extensively in a wide array of national and international scientific journals. He has authored numerous book chapters and proceedings and coedited one book entitled *Grazing Management: An Ecological Perspective* (Timber Press, 1991).

Dale W. Johnson is a professor of soils in the Department of Environmental and Resource Sciences, College of Agriculture, University of Nevada, Reno. His research interests are in soil chemistry and nutrient cycling. He has served on editorial boards and numerous advisory committees, given testimony to Senate Committees and the Department of Energy, been an invited speaker at several international conferences, authored over 200 publications (121 peer-reviewed), and coauthored or edited 3 books. He is a Fellow of the American Association for the Advancement of Science and the Soil Science Society of America. He received the Scientific Achievement Award from Environmental Sciences Division, Oak Ridge National Laboratory in 1983, Publication Awards from Martin Marietta Energy Systems in 1985 and 1987, Technical Achievement Award from Martin Marietta Energy Systems in 1986, the Dandini Medal of Science from the Desert Research Institute in 1993, and the Regent's Researcher Award from the University and Community College System of Nevada in 1999.

John D. Joslin, Jr. is an environmental scientist with Belowground Forest Research, which he founded in 1999 in Oak Ridge, Tennessee. He received a B.A. from Duke University, an M.S. from Cornell University, and a Master of Forestry degree from North Carolina State University. He received a Ph.D. in forest ecology and soil science from University of Missouri-Columbia in 1983, following a dissertation on the turnover of fine roots in a white oak stand. Most recently, his research has focused on the effects of climatic change on the development of forest root systems and the cycling of carbon in soils. He worked for the Tennessee Valley Authority at Oak Ridge National Laboratory and in Norris, Tennessee, for 17 years. Dr. Joslin has conducted research on the impacts of acid deposition on soil and stream-water chemistry, of aluminum on the development of tree roots, and of acidic cloud water deposition on high elevation forests. In addition to authoring over 40 peer-reviewed publications, Dr. Joslin served as chair of the North American Forest Soils Conference from 1993 to 1998, and is the current chair-elect of the Forest and Range Soils Section of the Soil Science Society of America.

Alan K. Knapp is a professor in the Division of Biology at Kansas State University. His research interests range from plant physiological ecology to ecosystem science with a primary focus on understanding the structure, function, and dynamics of grasslands. He has been involved in research at the Konza Prairie Long-Term Ecological Research program since 1982 and provided scientific leadership for the research program throughout the 1990s. His research has been published widely, and he coauthored *Grassland Dynamics: Long-Term Ecological Research in Tallgrass Prairie* (with John Briggs, David Hartnett, and Scott Collins, Oxford University Press, 1998).

Joseph R. McAuliffe is a research ecologist with the Desert Botanical Garden, a private, nonprofit natural history museum in Phoenix, Arizona. He earned his undergraduate degree in 1978 from the University of Nebraska, Lincoln and his doctorate in 1983 from the University of Montana. His research involves multidisciplinary studies of geomorphology, soils, and vegetation. In 1995 he received the W. S. Cooper Award from the Ecological Society of America for his research in the Sonoran Desert. His recent publications include refereed journal articles and book chapters that focus on the natural history of the Sonoran Desert.

James K. McCarron is a Ph.D. candidate at Kansas State University. His research examines the ecophysiological mechanisms of plant responses to anthropogenic disturbances and climate change. His current research investigates the mechanisms behind the expansion of C_3 shrubs into the C_4 dominated tallgrass prairie. He is funded by the National Science Foundation's Long-Term Ecological Research program at the Konza Prairie Biological Station and Kansas State University.

Richard Miller is a professor in the Department of Rangeland Resources at Oregon State University and stationed at the Eastern Oregon Agricultural Research Center. His research focuses on juniper woodland ecology, fire history, and the effects of past and current fire regimes on semiarid shrub-steppe and woodland communities. He has published in regional, national, and international journals that deal with ecology and management of natural resources and has authored several books.

Ronald P. Neilson is a bioclimatologist with the USDA Forest Service, Pacific Northwest Research Station, and a professor (courtesy) with the Department of Botany and Plant Pathology and the Department of Forest Science at Oregon State University. Dr. Neilson has focused on the theory, mechanisms, and simulation of vegetation distribution for nearly three decades. He received the Cooper Award from the Ecological Society of America for his research on oak distribution in the Rocky Mountain region. Dr. Neilson's MAPSS biogeography model and MC1 dynamic general vegetation model (DGVM) have contributed to national and global assessments by the Intergovernmental Panel on Climate Change (IPCC) and the U.S. Global Change Research Program. Dr. Neilson was the lead author for the forest sector on the IPCC's special report on *The Regional Impacts of Climate Change* and the convening lead author for an Annex to the Special Report on simulations of global vegetation redistribution under climate change. His current work extends into earth system modeling, landscape system modeling, and large-scale fire forecasting. Dr. Neilson received the Forest Service Chief's 1999 Honor Award for Superior Science. He received a B.A. in 1971 from the University of Oregon, an M.S. in 1975 from Portland State University, and a Ph.D. in 1981 from the University of Utah.

Elizabeth G. O'Neill is a research associate of the Environmental Sciences Division, Oak Ridge National Laboratory, Oak Ridge, Tennessee. She received B.A. and Ph.D. degrees from the University of Tennessee in the fields of ecology and botany, respectively. Dr. O'Neill's research focuses on the impacts of climatic change on root growth and litter dynamics of hardwood forest systems.

M. Keith Owens is a professor of rangeland ecology with the Texas Agricultural Experiment Station in Uvalde, Texas. His research focuses on the interactions of plants, animals, and the environment within savanna ecosystems. His major focal points are seed and seedling ecology of dominant trees and shrubs, water use of native shrubs and trees as they affect local water budgets, responses of savanna plants to growing and dormant season fires, and the spatial utilization of plant communities by herbivores. One of his current projects manipulates the amount, frequency, and season of precipitation in native savannas to determine how climate change may affect soil moisture recharge and seedling establishment. He has conducted research within the shrub communities of many western states and Alaska.

Keirith A. Snyder is currently a research scientist at the U.S. Department of Agriculture–Agricultural Research Service, Jornada Experimental Range, Las Cruces, New Mexico. Her current research focuses on the water relations of cold desert shrubs and the hydraulic redistribution and transpiration of water at night. Dr. Snyder earned her doctorate at the University of Arizona in 2001. Her dissertation research focused on the water use of dominant woody species in semiarid riparian systems, and understanding the relationship between groundwater depth, water source use, and plant ecophysiology. She received a master's degree in watershed management from the University of Arizona in 1995. Recent publications authored or coauthored by Dr. Snyder have explored (1) water source use of desert phreatophytes, (2) ecophysiological characteristics and survival of oak seedlings, and (3) variations in the functional rooting depth of experimentally manipulated mesquite.

Tony Svejcar is currently a research leader and rangeland scientist with USDA–ARS in Burns, Oregon. He has previously held research positions with ARS in Reno, Nevada, and El Reno, Oklahoma. Dr. Svejcar received B.S. and M.S. degrees from Colorado State University and a Ph.D. from Oregon State University. His current research interests include grazing impacts on rangelands, ecology and management of western juniper, carbon dynamics of rangelands, and climatic impacts on sagebrush steppe vegetation.

Donald E. Todd is a research technician at the Environmental Sciences Division, Oak Ridge National Laboratory, Oak Ridge, Tennessee. His research focuses on the response of the structure and function of eastern deciduous forests to changes in precipitation regimes.

David G. Williams is an associate professor in the Department of Renewable Resources at the University of Wyoming. Dr. Williams earned his doctorate in botany from Washington State University in 1992 and then spent two years as a postdoctoral fellow at the University of Utah. Dr. Williams joined the faculty at the University of Arizona in 1995. His current research investigates the role of plants in the hydrologic cycle within the context of global change in arid and semiarid ecosystems. Recent publications authored or coauthored by Dr. Williams have explored (1) interactive effects of atmospheric CO_2 enrichment and drought on C_4 grass photosynthesis, (2) intra- and interspecific variation in functional rooting depth in pinon-juniper woodlands across a monsoon precipitation gradient, and (3) hydraulic architecture and dynamics in the root systems of desert phreatophytes.

Index

Buxton Climate Change Impacts Laboratory, 74

calcic horizons, 11
California, 17–18, 34, 49, 60, 63, 67
Calliandra eriophylla, 14
CAM (Crassulacean acid metabolism) plants, 39
Canadian Climate Center (CGCMI), 3, 4
carbon: in leaf litter, 168, 175–77; root systems and, 38–39, 40
carbon dioxide, 3, 6, 91, 92, 129, 147; rainfall patterns and, 162–63; soil moisture and, 160, 161–62; and water uptake, 44–45
Cascade Mountains, 49
case studies, 7–8
Celtis reticulata, 34
Central Plains, 147, 148; vegetation and rainfall relationships in, 149–52
Central Valley (Calif.), 67
CENTURY biochemistry model, 53
CGCMI. *See* Canadian Climate Center
chaparral: interior, 16–18
Chihuahuan Desert, 17
Chrysothamnus nauseousus, 34
clay, 19, 20, 21
climate, 47, 170
climate change, 27, 147, 155; forests and, 164–66; modeling, 67–68; plant communities and, 4–5, 90–91; precipitation patterns and, 92–93, 164; and southwestern oak savannas, 129–30; vegetation distribution and, 22–23, 24; water uptake and, 44–45, 46
Coconino Plateau, 35–36
Collinsia parviflora, 96, 98
Colorado Plateau, 17, 49, 60
conifer forests: precipitation changes and, 66–67

cool season plants, 128, 148
cornbelt, 50, 51
Cornus florida, 165, 167, 168, 178
cottonwood: Frémont *(Populus fremontii),* 34, 40, 42, 44
creosotebush *(Larrea tridentata),* 20, 22
Crepis: C. *acuminata,* 96; C. *occidentalis,* 96
cyclonic storms, 49

decomposition: of leaf litter, 175, 176–77
Department of Energy, 165
desertification: Southwest, 58
deserts: air flow and, 56–57
DGVMs. *See* dynamic global vegetation models
dormancy: plant, 14
drive systems: in rainout-shelter design, 83–85
dropseed: sand *(Sporobolus cryptandrus),* 110
drought, 4, 56, 59, 69, 113, 118, 149, 161, 165, 173; and rangelands, 107, 109–10, 115–17, 123–26; root growth and, 120–21; and Walker Branch TDE, 17–72; and xylem water potential, 121–22
drought deciduousness, 14
dustbowl era: on Central Plains, 149
dynamic global vegetation models (DGVMs), 23–24, 26, 187–88

Eastern Broadleaf Forest Province, 165
eastern deciduous forest, 8
ecosystems, 3; precipitation changes and, 186–87; precipitation regimes and, 5–6, 185–86
ecotones, 64, 65(fig.), 67, 68, 149
electricity: rainout shelter design and, 83–84

El Niño Southern Oscillation, 69, 129, 182

Eragrostis intermedia, 20

evapotranspiration: rangeland water budgets and, 123, 124

fencing: in rainout shelter experiments, 88

fescue: sixweeks *(Festuca octoflora),* 110

fire, 16, 142

fire models, 53, 56

Flint Hills (Kan.), 148, 150–51, 161. *See also* Konza Prairie Biological Station

Florida, 50

foliage, 167–68

forbs, 122; on Konza Prairie, 149, 155, 160. *See also* herbaceous plants

forests: and precipitation patterns, 66–67, 164–65, 170; TDE in, 165–78

Fort Huachuca Military Reservation, 128

Fort Keogh Livestock and Range Research Laboratory: precipitation experiment on, 108–26

frost: oak distribution and, 60, 64

frost-free period: in Prairie Peninsula, 55–56

Garden Canyon, 29, 128–29; precipitation experiment at, 130–42

gas exchange: root systems and, 29

GCMs. *See* general circulation models

general circulation models (GCMs), 3–4, 6–7, 23, 187

geology, 7; and vegetation distribution, 17–19

geomorphology: soils and, 24–26

global warming, 4, 47; vegetative responses to, 67–68

grama *(Bouteloua),* 21(fig.), 108; black *(B. eriopoda),* 20, 117–20, 121; blue *(B. gracilis),* 14, 110, 149; sideoats *(B. curtipendula),* 20, 149

granite: weathered, 17–18

grasses, 16, 59, 67, 127, 129, 137, 148, 149; Konza Prairie, 149, 155, 160, 161; livestock and, 20–22; rangeland, 114–26; root systems of, 28–29; soil formation and, 19–20; water use by, 12, 150

grasslands, 57, 58, 144, 147; rainfall manipulation experiments on, 152–63; rainfall patterns and, 49, 150–52, 156–60, 162–63; semidesert, 14, 16–17, 18–22. *See also* rangeland

grazing, 16, 107, 109, 113; impacts of, 114, 115–16; nutrient losses from, 124–25; and perennial grasses, 20–22; and tiller growth, 117–18, 120

Great Basin, 60, 67, 68, 90; precipitation patterns in, 92–93; soil water-plant community experiment in, 93–106

Great Plains, 50; precipitation models for, 4, 54–56; rainfall patterns on, 57, 155; rangeland productivity on, 108, 160–61. *See also* Northern Great Plains

groundwater: water uptake from, 34, 40, 42(fig.), 46

growing season: and precipitation, 35–36, 170

Gulf Coast, 4

Gulf of California, 57

Gulf of Mexico, 49, 54, 57

Gutierrezia spp., 14, 32

hackberry: netleaf *(Celtis reticulata),* 34

HADCM2 (Hadley Centre, U.K. Meteorological Office), 3, 4, 6, 129

rangeland, 107; expansion and contraction of, 59–60; germinable seedbanks on, 122–23; herbage biomass on, 114–17; productivity of, 91, 108–26

recharge: precipitation and, 10–11

RegCM. *See* regional climate model

regional climate model (RegCM), 6

riparian areas, 34; water uptake in, 42, 44

Rocky Mountains, 48, 49, 57, 64, 67

root cavitation, 32

root growth: forest, 167, 174–75; rangeland, 108, 111, 120–21

root systems, 28–29, 108, 181; biomass of, 136, 137(fig.); hydraulic redistribution and, 30–31; water uptake of, 32–46; and water use, 13–15, 16, 31–32; and weathered rock, 17–18

runoff, 76, 112–13, 125

sagebrush *(Artemisia)*: big *(A. tridentata),* 34, 74; Wyoming big *(A. tridentata* subsp. *wyomingensis),* 93, 96, 98, 104, 105

sagebrush steppe community, 8; precipitation timing and, 102–3

Sahara Desert, 57

Salix gooddingii, 34, 40, 42

San Pedro River, 34, 44(fig.)

Santa Catalina Mountains, 21(fig.)

Santa Rita Experimental Range, 21

saplings, 36, 165

savannas: and precipitation changes, 66–67; soil resource partitioning and, 144–45; southwestern oak, 128, 129–45

Schizachyrium scoparium, 149

seasonality, 3, 183; of precipitation, 66–71, 90–91, 148, 178, 182; soil moisture and, 156–57

sedges, 148

seed banks: germinable, 108, 111–12, 122–23

seedlings, 15, 36, 133; emergence and recruitment of, 134–36, 141–43, 144–45

semidesert grasslands, 14, 15; distribution of, 16–17; geology and, 18–19; grasses in, 19–20; livestock grazing and, 20–22

shortgrasses, 110

shortgrass prairie, 14

shrubland: Southwest, 58

Shrubland Subtropical Xeromorphic (SSX), 60, 61(fig.), 68

shrubs, 14, 74, 149

Sierra Madre Occidental, Bermuda High and, 57

Sierra Nevada, 49

Sitanion hystrix, 96, 98

snakeweed *(Gutierrezia* spp.), 14, 32

snow, 80

soil(s), 7, 8, 9, 22, 23, 93, 150, 165; basalts and, 18–19; characteristics of, 162, 168–69, 180–81; geomorphology and, 24–25; hydrology of, 25–26; and plant distribution, 19–20, 21(fig.); water percolation in, 10–11

soil moisture, soil water, 6, 9, 30, 54, 136, 181–82, 183; grasslands and, 150–52, 154(fig.); in Northern Great Basin Experimental Range, 93, 94–95; nutrient availability and, 161–62; partitioning of, 127–28, 143–44; and plant communities, 93–106; plant productivity and, 157–60; rainfall patterning and, 156–57; rain shelters and, 104–5; in rangelands, 113–14, 123–24; spatial and temporal distribution of, 10–13; vegetative responses to, 15–20; in Walker Branch Watershed, 166–67

soil water potential (SWP), 166

solar radiation, 166, 170; rainout shelter design and, 81–82

Solidago canadensis, 155, 160, 161

Sonoran Desert, 15, 16–17, 39

Sorghastrum nutans, 149

Southeast (U.S.), 4, 50, 56, 57

southern Rockies, 4

Southwest (U.S.), 10, 64–66; precipitation models in, 4, 6; precipitation patterns in, 49, 57–58, 69–70; temporal precipitation patterns in, 58–63

Sporobolus cryptandrus, 110

sprangletop: green *(Leptochloa dubia),* 20

spring, 100, 103–4, 105

squirreltail *(Sitanion hystrix),* 96, 98

SSX. *See* Shrubland Subtropical Xeromorphic

Stipa: S. comata, 110; *S. thurberiana,* 93, 96, 98

subshrubs, 14

Subtropical jet stream, 49

succulents, 39

Sulphur Springs Valley, 19

summer, 100; precipitation during, 49, 54, 63, 67, 69, 70, 91; soil moisture partitioning, 127–28; water percolation in, 10–11; water uptake in, 34–35, 40–42, 43(fig.)

switchgrass *(Panicum virgatum),* 149

SWP. *See* soil water potential

tallgrass prairie, 148, 152

Tamarix ramoisissima, 34

tanglehead *(Heteropogon contortus),* 20

TDE. *See* throughfall displacement experiment

temperature, 3, 68, 91; soil, 96, 100–101, 103(fig.), 104–5

Texas, 17, 22, 57, 88

throughfall displacement experiment (TDE): on Walker Branch Watershed, 165–79

tiller growth: dynamics of, 117–20; measuring, 110–12

tobosa *(Pleuraphis mutica),* 14, 20, 22

topography, 17

Trachypogon montufari, 128, 134, 137, 143, 144

trees, 34, 56, 66, 149, 167

trigger mechanisms: in rainout-shelter design, 85–86

Tucson, 57

two-layer hypothesis, 58, 69–70, 127–28, 148

understory, 53

United Kingdom, 3, 74

U.K. Meteorological Office, 6

Utah, 40, 68

vegetation, 11, 48, 52(fig.); Central Plains, 149–52; global warming and, 67–68; measuring biomass of, 95–96; modeling distribution of, 66–67; precipitation and, 48, 63–64, 66–67, 69–70; and soil conditions, 9–10; Southwest, 58–66

vegetation zones: precipitation and, 63–66

Verde Valley, 18

Walker Branch Throughfall Displacement Experiment (TDE), 164; description of, 165–69; results of, 170–79

Walker Branch Watershed: TDE on, 165–79

Walnut Gulch Experimental Watershed, 10
warm-season plants, 128, 148, 149
water: plant use and, 13–15, 181; rainout-shelter design and, 86–87; recharge of, 10–11; and soils, 9–10; woody plant use of, 29, 32–46
water budgets: rangeland, 123–24
water stress: Walker Branch TDE, 171–72
water yield: in rangeland experiment, 112–13
wedge hypothesis, 64, 66, 68
wheatgrass: *Agropyron* sp., 108; bluebunch *(Pseudorogenria spicata),* 93; western *(Pascopyrum smithii),* 110, 118, 120, 121
Willamette Valley, 67

willow: Goodding *(Salix gooddingii),* 34, 40, 42; seep *(Baccharis glutinosa),* 34
wind: and rainout-shelter design, 79–80
winter, 49, 102, 104, 105; precipitation patterns in, 59, 67, 69
woody overstory, 53
woody plants, 17, 66, 67, 127; establishment of, 69–70; livestock grazing and, 20, 21–22; Prairie Peninsula, 54–55; root systems of, 28, 29, 31–32; seedling recruitment in, 141–42; water uptake by, 32–46; water use by, 14, 15, 16

xylem water potential, 108, 111, 121–22